Psyche's Palace

PSYCHE'S PALACE

HOW THE BRAIN GENERATES THE LIGHT OF THE SOUL

Essays on the Biological Origins of the Sentient Being
and the Theory of Neurobioluminescence
in the Evolution of Sensorium Consciousness

David Aaron Holmes

© Copyright 2008 David Aaron Holmes

The Library of Consciousness

All rights reserved

ISBN: 978-0-6151-6411-3

Cover design and illustrations by David Aaron Holmes

Photograph (p.271) by Kenneth Holmes

This book is listed in major international bibliographic databases and is available to booksellers worldwide.

To order single copies of this book on-line, please go to:
http://www.lulu.com/content/1173020

To contact the author, please e-mail him at:
DavidHolmes@psychespalace.com

To visit his web site, please go to:
http://www.psychespalace.com

Several early drafts of this book, variously entitled: *A Play of Light: The Luminous Theater of the Mind—Unveiling the Sensuous Brain* (© Copyright 2005 David Aaron Holmes, The Library of Consciousness, ISBN: 978-1-4116-4116-7 [Library of Congress Registration Number TXu1-238-745]), *The Bioluminescent Brain: Theater of the Enlightened Mind* (© Copyright 2006 David Aaron Holmes, The Library of Consciousness) and *Psyche's Palace: How the Brain Generates the Light of the Soul* (© Copyright 2007 David Aaron Holmes, The Library of Consciousness) were published in small runs. This 2008 edition is the final and definitive version of this book.
Les jeux sont faits. Rien ne va plus.

In deep gratitude
to my parents, George and Jane Holmes,
to my cousin Elizabeth Vilensky,
to my uncle Kenneth Holmes,
and to my friends Douglas Walters,
Bill Blackburn, Allen Holt, Arthur Peña,
Stephen Whitmarsh, and Julie Minton,
for their invaluable help and support.

CONTENTS

INTRODUCTION	x
Chapter One **A PLAY OF LIGHT**	1
Chapter Two **THE GLOBE THEATER**	11
Chapter Three **THE MYTH OF CUPID AND PSYCHE**	19
Chapter Four **NEUROBIOLUMINESCENCE**	29
Chapter Five **CONCEPTUAL FRAMES**	43
Chapter Six **EGOGENESIS**	52
Chapter Seven **AS WITHOUT, SO WITHIN**	64
Chapter Eight **RUNNING FOR COVER**	76
Chapter Nine **MOOD RINGS**	93
Chapter Ten **THE THREE POISONS**	106
Chapter Eleven **MISTAKING THE MAP FOR THE TERRITORY**	117
Chapter Twelve **NOTHING DOING**	131
Chapter Thirteen **EMBRACING THE GENTLE MOTHER**	141

Chapter Fourteen
FURTHER CONJECTURE 151

Chapter Fifteen
THE STREAM OF CONSCIOUSNESS 168

Chapter Sixteen
CHILD'S PLAY 178

Chapter Seventeen
ONCE MORE, WITH FEELING 193

Chapter Eighteen
HIDDEN TREASURE 214

Chapter Nineteen
BIRTH-BREATH-DEATH 235

Chapter Twenty
BOTH SIDES NOW 251

AFTERWORD 255

INDEX 257

(Note: a one-page summary follows each chapter)

ILLUSTRATIONS

p. xii —***The Miracle of the Lobes and Fissures***— At the moment of fully receiving the unexpected present, the black box of the brain opens in exploded view. What is inside? *No Thing!* Deep down you know that this is just what you have always wanted. Upon the turbulent surface of the very ocean where Psyche was born appears this overwhelming wave of realization—the violent surf we one day learn to ride with grace and ease.

p. 10 —***The Globe Theater***— Meandering through the primeval forest, tucked deep within his sensorium sphere, the daydreamer—flesh brain on walking sticks—conjures up his own real world in deference to and imitation of the *real* real world outside. We know we *are* only because we are *inside* of something else that is clearly (visibly, audibly, and palpably) real. Every sphere must have a center, and here, within this particular sphere, *I* must be *it*.

p. 28 —***Neurobioluminescence***— Three simultaneous fountains of light spring up from the brainstem in each moment of embodied consciousness—the fountain of the aeon, the fountain of the lifetime, and the fountain of the fresh moment. The particular lineage of ancestors illustrated here culminates in the horned atlantes that now shoulder their great, glowing sphere of illumination. Chain mail rings of neuronal oscillators bring harmony, order, rationality, and beauty to the flurry of neural chaos—and much welcome chaos to the tyranny of rigid order.

p. 92 —***Mood Rings***— We exist as a vibrating, microcosmic image of the macrocosmic, vibrating universe. Pure color reveals the smooth, periodic oscillation of the electromagnetic field; pure musical tone, the smooth propagation of waves of air pressure. The minute and meaningful periodic vibrations that tickle the washboards of our fingerprints reveal intimate surface details of the objects we touch. Our senses can only make sense of what repeats. Rational relationships of temporal events construct the dimension of time, as echoed in the harmonic overtones, chords, and rhythms of music. Illustrated here are the first seven vibrational modes of the top and back plates of a violin and a series of Lissajous curves that describe progressively more complex harmonic motion descending into the chaos of the natural world. Other periodic forms float by in analytical boxes.

p. 116 —***Mistaking the Map for the Territory***— The befuddled, knuckle-dragging homunculus gawks from the sidelines at the fantastic developments. He is the laughingstock of the unreflective Cartesian dualist, a distorted caricature trotted out to frighten young neurobiology students and dissuade them from their preposterous soul-searching within the incontestably subcelestial brain. *David* provides a nobler self-image for our newlywed tourist, who glances up from his brochure just long enough to take in the full panorama of the *real* Niagara Falls, displayed in all its glory upon the backside of his brain.

p. 140 —***Embracing the Gentle Mother***— The carpet weavers crouch over the rustic loom and whip their agile, dendritic fingers through the warp strings. They tie new knots of colorful silken yarn to generate infinite patterns of sensuous pleasure and to code the intricate processes of mind and brain. The surface of the cerebral cortex is lined with a thin, transparent membrane *(pia mater,* the gentle mother) that physically separates but energetically *unites* the cortex with the crystal-clear sea of cerebral spinal fluid that surrounds it—this in order to optimize the dazzling display. The two hookah smokers appearing above and below are one and the same; they are you and your wild hallucination.

INTRODUCTION

This book offers what I hope you will find to be a refreshingly plain and lucid theory of consciousness—one that satisfies the rational scientific impulse yet accords with the most profound and ancient spiritual intuitions. It is not an attempt to persuade you toward or dissuade you from any particular view of the divine cosmic order. Nevertheless, the scope of this theory is vast—as it must be, considering the immensity of the subject matter. The theory calls for a thorough reexamination of the brain's neuroanatomical structures and processes, employing a fresh conceptual vision. It delves deep into our own evolutionary history—to the very origins of sentient existence—exposing the archaic roots of our most problematic human emotions and behaviors and confirming how difficult will be the task of releasing ourselves from their tenacious grip.

The theory is predicated upon a single metaphysical postulate—that *light* itself is aware; awareness itself is *light*—that the quantum field is brimming with an intrinsic but initially undeveloped proto-awareness, from which our own terrestrial forms of embodied consciousness are fashioned by specific evolutionary processes. This is where the theoretical model first emerges into view from behind the veil of the Great Mystery of the union of spirit and light—or in the language of modern physics: the inseparability of the *observer* and the quantum event. From here a much sturdier and more orthodox chain of logic and scientific inference will guide us toward an understanding of what embodied consciousness actually is and how it came to be.

Although microscopic in size, neurons within the brains of animals generate inordinately strong electrical voltages when they fire. The small perturbations of the electromagnetic field that accompany these firings are, I contend, the source and conduit of the *Light* of conscious awareness. The conscious mind manifests as an intricately patterned, *pixellated* quantum field created by the coordinated firings of the hundred billion neurons of the cerebral cortex. By means of countless intricate "neurobioluminescent" effects, *Light* itself is shaped by the brain into the vivid three-dimensional presentations of sight, sound, smell, taste, and touch that form the *sensorium* within which we indirectly experience our existence.

Surrendering to the ineluctable force of evolution, *Light* is linked to mind, and mind to matter, as the brain links billions of brightly firing neurons together into elaborate circular chains. These complex oscillators combine to form a vibrant, multisensory *neural resonance chamber*. Its stock repertoire of rousing tunes and ominous chord progressions alternately motivate and dissuade the inwardly luminous sentient being—now endowed with a full complement of emotions—toward the desirable and away from the aversive.

From the *Unborn Light* to mind to matter—to the dense physical body with its primal impulses of desire and aversion—we trace the path of spirit from an initial state of inherent freedom to one of virtual enslavement. Implicit within *the Theory of Neurobioluminescence in the Evolution of Sensorium Consciousness*, however, is the realization that nothing can ever separate us from what we fundamentally *are*—the free and blissful *Light of Awareness*. All that keeps us in this apparent bondage is the *appearance* of bondage itself—which is termed "ignorance" in Buddhist philosophy—a mistaken identification with this mortal body and its compelling presentation of a solitary and constricted, but ultimately illusory *self*.

Christians celebrate the Transfiguration of Christ as the culmination of his teaching, wherein he revealed his divine inner *Light* to three of his disciples upon a high mountaintop. "His face did shine as the sun, and his garments became white as light." Saints and mystics have been described as emanating halos of golden light, and some people claim to be able to see colored auras surrounding all living beings. Practitioners of Tibetan sleep yoga nightly meld their waking consciousness with the primordial *Light of Nonduality* that rises from their brainstems like silent stars upon their deep and dreamless delta sleep. *"Mehr Licht!"* ("More light!")—thus cried out an ecstatic Goethe with his final breath. In so many cultural and religious traditions, the vocabulary of spiritual attainment makes reference to an increasingly brilliant clarity: an "enlightenment" or "awakening" or "illumination." Perhaps the most rational explanation for this universal insight is that it is quite literally true. We actually are beings of *Light*—right now—just as we are.

CHAPTER ONE

A PLAY OF LIGHT

It was foggy out, but a bright, luminous fog that promised sunshine would break through within an hour or so. I was sitting at the kitchen table, daydreaming—waiting for that second cup of coffee to kick in. I could easily squander a whole morning like that in those days, mulling over some existential conundrum or another. Like a face-off with a classic Zen koan, it was just another way to wrestle the analytical mind into exhausted submission, in the hope that something fresh and valuable would emerge. But the real answers seemed forever elusive.

Somewhere in the mental mix were the perennial questions about the nature of consciousness: what is the relationship of the mind to the brain, and what, if anything, is the soul? Most of these musings were hard to square with the sterile, reductionist paradigm then current—that the brain was nothing more than a biological computer and a clumsy, unreliable one at that—destined soon to be displaced by the bright and shiny, super-efficient supercomputers of the future.

Scientists in general and neurobiologists in particular seemed curiously incurious about the subject of consciousness, unwilling to tackle the problem or even to admit that there was one. René Descartes, three centuries before, had set the course of Western science by hacking the living universe into the two separate spheres of mind and matter, graciously relegating the question of consciousness to philosophers and priests. As far as any self-respecting neurobiologist was concerned, only that which could be examined and measured and verified with brain probes seemed worthy of scientific consideration. As for the rest—the inner experience—was it really even happening?

This was 1980; the neurobiological sciences were still firmly in the grip of *epiphenomenalism*—the doctrine that consciousness is merely an "epiphenomenon" (an incidental byproduct) of the brain's physiological processes. Epiphenomenalism's peculiar and scientifically untenable postulate is that the conscious "self" is simply an illusion—immaterial in both senses of the word—a secondary phenomenon of no real importance, having no power to affect either the brain or the body.

In any event, I didn't set out that morning to deal a death blow to Cartesian dualism or to solve the mind/body problem. I was simply enjoying a cup of coffee, minding my own business, looking out the kitchen window at the morning fog.

Slowly, out of the gray haze, there came to my mind an image. It was the simplest, most innocuous thing, like a tiny seed taking root in the moist, mushy soil of my brain—teeming as it was in those days with a great deal of speculation, but as yet unencumbered by many actual facts. This image persisted and grew and gradually elaborated into a more complete picture and a hunch, and then a hypothesis and a theory; and before I got up from the table that morning I found I had a full-fledged *Weltanschauung* (I'd use the English term "worldview," but it doesn't quite capture the Wagnerian feel of the experience) ricocheting between the walls of my skull. It was a delightful, giddy feeling, like a secret window had opened up to a new dimension.

What I saw in my mind's eye was a circle of lights, maybe eight lights in all, each blinking on and off in succession—a beautiful, rotating pattern of glowing lights. All at once I realized that these were brain cells—neurons—that I was visualizing: eight neurons connected in a ring, one firing into the next, the last firing into the first—creating this illusory movement of light in the same way that the blinking of theatrical chase lights causes glowing patterns to appear to revolve around a theater marquee.

A PLAY OF LIGHT

I knew that neurons did not actually visibly *glow* when they fired... or *did* they in some sense? A neuron does produce a rather spectacular, seemingly extravagant electrical voltage when it fires, as an explosive flow of positively charged sodium ions streams through voltage-gated channels in the cell membrane. Any such displacements of charged atoms—including the electric "currents" that flow along the neuron's axon and dendrites—induce corresponding changes in the local flux density of the magnetic field. These complex interactions between the electric and magnetic fields are mediated by real and virtual *photons*, the quantum particles responsible for the propagation of light and other forms of electromagnetic radiation. Therefore, perhaps the ostensibly "invisible" electromagnetic pulse generated by a firing neuron is in actuality *visible*—discernible to the indwelling consciousness as a minute, localized pulse of *light!*

Then it dawned on me that the steady pattern of firing in this ring of eight neurons would be of a predictable frequency, completing a circuit in a specific time interval: say, eight one-hundredths of a second. In other words, the ring of neurons formed a simple oscillating resonator. By joining together multiple, identical units of eight such neurons, the brain could easily create miniature timing mechanisms, groups of oscillators all operating synchronously.

The image became more elaborate, with rings of neurons of all sizes and shapes interconnecting with other rings, forming more complicated patterns of interaction—some rings working in harmony with one another, some out of sync with and therefore dampening the activity of others. For example, the ring of eight neurons could operate in conjunction with a ring of four such neurons in a tighter circle. That ring of four neurons would complete the shorter circuit in half the time and could therefore be used to maintain a frequency twice that of the ring of eight. A ring of sixteen neurons would take twice as long, sixteen one-hundredths of a second in this case, to complete a circuit and could therefore oscillate at half the frequency.

In musical terms this would represent an octave above and an octave below the original frequency. As a musician, I recalled that all of the basic chordal structures and rhythmic patterns that make tonal music so immediately accessible (and universally enjoyable) are invariably formed from simple ratios, fractional relationships of whole numbers. A ring of five neurons with a ring of six neurons could resonate a minor third (with a pitch ratio of 6:5). A ring of four neurons with a ring of five neurons could resonate a major third (with a pitch ratio of 5:4).

It struck me that all of this oscillation imagery I was seeing in my mind made sense to me because of my familiarity with music, musical instrument construction, and acoustics: that complex sound waves can be produced—or reproduced—by combining the effects of many independent oscillators. But this was a *visual* image, and the medium that was resonating here was not sound, but *light!*

I realized that these resonators were basic building blocks that could be cunningly arranged by the craftsmanlike hand of natural selection through the course of many hundreds of millions of years of evolution, developing and perfecting an elaborate multisensory resonator of visual, auditory, and other sense data. This meant that one of the main functions of the brain was to produce this *sensorium*, the vast sound and light extravaganza, for the benefit and entertainment of the indwelling soul. It didn't occur to me at that time what the darker, existential ramifications of such an intimate relationship of body and soul might mean philosophically. It was simply a welcome relief from the confines of the soul-deadening "brain-as-computer" model that was so pervasive in the scientific literature of the day.

What *these* neurons were doing seemed quite different from the type of active information processing one would expect to see if they were indeed the neurobiological analogues of computer circuits. The *activity* of these neurons was completely *passive*. They behaved collectively more like a vibrating medium than like individual intelligent agents.

And all were receiving streams of input data from sense organs— eyes and ears translating waves of light and sound into patterns of neural resonance. I imagined all sorts of configurations: rings in which the labor was divided between those neurons with a particularly "bright" electromagnetic discharge—which actually formed part of the plane of presentation—and the rest, which were hidden below, like members of an ancient drumming circle, whose job it was to keep a steady beat.

Yet it wasn't just the mechanistic oscillations of those fleshly neurons that seemed so amazing to me. However unimaginably intricate may be the meshwork of axons and dendrites of the hundred billion neurons that make up the human brain, when all is said and done, those neurons are simply *cells*, progeny of the same fertilized egg that also produced the cells of the liver, the heart, and the spleen. Neurons likewise exist entirely within the material world, the earthbound spawn of our genes and DNA. How could *they* then be the conduit to the subtle and mysterious realms of consciousness?

What became clear to me on that day was that there was *another* medium oscillating simultaneously in absolute synchrony, creating a perfect three-dimensional mapping of every pulse in the chain of neuronal firings—and it wasn't an "epiphenomenon." It was the main event! The medium I'm referring to is the *electromagnetic field*, a constant but ever shifting feature of the living brain. Although this field is also a completely objective reality, made tangible in electroencephalogram printouts or indirectly visible in the colorful displays of functional magnetic resonance imaging devices, we are nonetheless broaching a much more mysterious and incorporeal realm of energy and light.

The images produced by these machines are only crude approximations of the actual electromagnetic field. No one has yet seen an accurate image of the true form of this neurobiologically produced electromagnetic field in the kind of microscopic detail that I'm trying to describe here. Actually, you *have* seen such an image. In fact you are seeing one at this very moment as you read this page! What I'm suggesting is that your own consciousness—your mind, your *soul*, if you will—is a complex and highly organized entity, as intricately structured as the brain itself. The mind has evolved in complexity in concert with the brain, for the mind's primary task is to receive and experience the resonant images that are continuously produced upon the brain's surfaces. The mind operates quite literally as *the* real-time functional electromagnetic imaging device *par excellence*.

I had read enough of the Christian and Sufi mystics and had practiced Zen meditation long enough to know that light was not simply a *metaphor* for consciousness. Time and again in the sacred writings of nearly all religious traditions *Light* is referred to as the very stuff and substance of the mind and of the soul. Somehow, I surmised, this *Light*, this *neurobioluminescent quantum field*, must be the bridge between the physical and the metaphysical, the direct link between the material and the spiritual worlds.

Disturbances of the electromagnetic field involve the release and absorption of photons and other elementary particles. These photons' energy states may or may not place them within the narrow band of wavelengths corresponding to the spectrum of *visible* light, but it seemed entirely possible that an indwelling consciousness could be sensitive to a much broader spectrum of electromagnetic energy and would perceive it all as various forms of light (or, for nonvisual data, as patterned vibrations of some other regions of the electromagnetic spectrum where sound, smell, taste, and touch are displayed).

And realizing all of this, I thought to myself, "Maybe the brain isn't a computer at all! Perhaps it's more like a Stradivarius violin!"—one capable of resonating all of the sense modalities into subtle patterns of light. Maybe the neurons of the visual cortex, for example, aren't all arranged in hectic "telecommunication networks" with one another, their axons and dendrites sending a flurry of relevant data bits back and forth in order to "figure out" the world conceptually. What if the bright display of a firing neuron is an end in itself—a *pixel* (like one of the tiny dots of phosphorus on a color television screen)? What if the important *information* being transmitted is simply the appearance and disappearance of this speck of *light*, which in conjunction with the well-timed firings of billions of other neurons forms the panoramic image that we experience in our mind's eye as a three-dimensional visual field, a detailed reconstruction of reality meticulously re-created from the patterns of light received by the eyes?

Perhaps all of the sense modalities are perceived similarly as shifting patterns in the electromagnetic field. In that case—as odd as it sounds—the brain should really be viewed more like a *gland* that *secretes* organized patterns of light—perceived by the soul within what was always presumed to be the dark confines of the skull. Maybe the firing of neurons is the closest approach biological matter can make to the realm of the spirit; and it is *light*, electromagnetic radiation, that is the "substance" of this interaction of brain and mind, body and soul.

Here was a model that matched more closely how it felt to me to be a human being. I relished the thought that I wasn't a computer after all; I was actually a finely tuned musical instrument! There was a reason why harmony felt harmonious—and chaos chaotic—why my mind and spirit responded so joyously to a Bach concerto or the complex geometric patterns of a Persian carpet. It was all about ratios, proportions, the Golden Mean, relationships, rhythms, and major and minor chords.

Analogue, not digital! My experience of consciousness wasn't the bottom line of a complex, binary calculation of ones and zeros. Nor did it in any way resemble the instantaneous parallel processing of all permutations of logical trajectories that gave the insufferable illusion of genius to the computer chess master, *Deep Blue*.

In the twenty-seven years since that foggy morning breakthrough I've grown to appreciate the intricate computational data processing that actually does go on in the brain, and my either/or position on the computer/resonator issue has given way to a both-and. In fact these two metaphors hardly begin to express the actual complexity of the systems that sustain the inner experience of an embodied sentient being. But this insight was a starting point and, at the time, a great source of liberation in my thinking.

This isn't exactly neuroscience—not yet, anyway. It's simply a *Gedankenexperiment*, consisting so far almost entirely of unbridled speculation. But the mind is a wild elephant and a mystery, and it may not give up all of its secrets to the sober scientists who are in charge of its investigation. Are they really even looking for that deep, subtle realm of consciousness that is the source of all beauty, wonder, and intuition? Is it really a more rational and objective stance to posit the *absence* of a spiritual dimension rather than to acknowledge, after a moment of reluctant introspection, its indisputable *presence*?

Neuroscientists skirt these deeper spiritual issues by referring to this conundrum as the *binding problem*: "How are the attributes of an object, which are analyzed separately in physiologically distinct areas of the brain, bound together?" Ever leery of invoking anything resembling the *homunculus*, "the little man inside the head," or, God forbid, the *soul* as a causative unifying agent, brain researchers and cyberneticists constrain themselves in their search to the tangible neuronal structures and their associated physiological activity—what they seek is a form of consciousness entirely *of* the neurons, *by* the neurons and *for* the neurons.

To my mind, it's as if a race of robotic aliens were to discover an automobile on an uninhabited earth and puzzle over how it ran itself.

Studying the odd positioning of the pedals, steering wheel, and gear shift, the alien scientists may entirely ignore the well-worn upholstery and muddy carpets, the cigarette lighter, the glove box, and the vanity light—obvious and telltale signs of a human presence—and go on puzzling forever over the "binding problem."

What binds these aspects of mind together is not to be found in the physical maze of neurons, axons, and dendrites. Mind is the medium of *Light* invisible, the fluid electromagnetic field of energy that surrounds and subsumes all of brain space—a field that by its very nature is intimately bound, but only unto itself, and infinitely unbounded.

Neuroscience has been hamstrung by the burdensome legacy of Descartes: his dissection of reality into *res cogitans* and *res extensa*—"the thinking thing" (mind or spirit or soul) and "the extended thing" (physical matter, forces, and energies)—the first the purview of religion and philosophy, the second of science. Still to this day most neurobiologists would cringe at the use of the word soul written anywhere in the vicinity of the word brain (or if they themselves should use the term, would likely hold it at arms' distance with punctuation's equivalent of a pair of latex gloves, implying that it should be read, with a marked tone of disdain, as—*the "soul"*—). I refuse to put the word in scare quotes. I don't disown it; nor do I, as a staunch Buddhist, take myself to *be* the soul. But I certainly do embrace it as my most prized possession.

We all approach this study with some established prejudices. Are we willing to ease up a bit in this—in what is perhaps the most important investigation we humans will ever attempt: a final unveiling of the secret life of consciousness? We must carefully consider every clue, from whatever far-flung field or discipline it may come.

Remember the fable and its moral: that the part of the elephant that you are holding on to, as we all grope about in the dark tent, whether it feels like a fan, a rope, a snake, a spear, a wall, or the trunk of a tree, is only one aspect of a vast reality—the infinite mystery of the mind. Neuroscientists have their way of approaching the enigma. Mystics, gifted with insight, also have vital information to share. This is everyone's story. Not only the poets and philosophers, artists and musicians, but every man, woman, and child—all those who secretly yearn to know what it truly means to be alive and awake—are wholeheartedly welcomed to join in *this* conversation.

A PLAY OF LIGHT

A SUMMARY OF CHAPTER ONE

This new theory of consciousness was born from the insight that a class of neurons in the brain may actually be functioning in the same manner as the colored pixels (the tiny phosphorescent dots) on a television screen. Individually, their firings produce only infinitesimal electromagnetic bursts of an inwardly visible light, perhaps in an assortment of wavelengths (an array of colors). In the aggregate, however, this coherent, spatially extended, fine-grained quantum electromagnetic field is directly experienced by and as the Light of consciousness. These presentation neurons are the natural conduit between the body (matter) and the mind (an energetic field of Light).

Presentation neurons can be arranged in circular rings, one firing into the next, like colored lights around a theater marquee. Since the constituent neurons can be made to fire at predictable rates, these rings behave like simple oscillators or resonators. Paired rings of various sizes are able to resonate and display simple ratios. In musical terms, for example, a ring of five neurons with a ring of six neurons forms a pitch ratio of 6:5 and can resonate—and therefore present to consciousness—an interval of a minor third. The neural resonance chamber can be compared to a violin—one that produces and resonates tremendously complex patterns of Light. The brain serves as an aesthetic device by means of which consciousness determines what is pleasing (desirable) and displeasing (undesirable).

These "pixels" and resonators are the building blocks used by natural selection to improve and refine the presentation. In due course they evolve into a platform that supports and presents the vast sound and light extravaganza that now surrounds you: the sensorium.

The electromagnetic field produced by the brain is minutely detailed, and it is precisely at this level of detail that we directly experience sights, sounds, and bodily sensations. Even high resolution images of the electromagnetic field surrounding the cerebral cortex give no hint as to how intricately the field is actually arranged for presentation at the microscopic level.

CHAPTER TWO

THE GLOBE THEATER

The heart pumps blood. The lungs breathe air. *The brain presents reality.* It is as startlingly simple as that. The heart and lungs make no attempt to conceal their mechanical function from the observant anatomist; nor does the brain, truth be told—although it requires somewhat keener sight and insight to discern this truth. The brain's long-secret function is brought to dazzling light when the question is properly posed: just what is a cerebral cortex neuron actually doing when it fires?

The *second* most obvious thing that a cerebral cortex neuron is doing when it fires is sending an impulse along its axon to the axon terminal, releasing neurotransmitters through the presynaptic membrane into the synaptic cleft, exciting or inhibiting the postsynaptic neuron—to all appearances functioning as a rudimentary circuit, a logic gate, the atomic unit of the *biocomputational brain.* But the *first* most obvious thing that a firing cerebral cortex neuron is doing is *firing.* As far as consciousness is concerned, the neuron's firing is "the be-all and the end-all"; the intricate spaciotemporal patterns of neuronal firings within the sensory cortices directly generate all objects of awareness. Consciousness perceives the firing neuron as a single, illuminated pixel with a particular luminosity at a particular location in brain space. Consciousness *is* the totality of these minute fluctuations of the cerebral quantum electromagnetic field.

Your experience of the world takes place in the *sensorium*, a subjectively centered, circumambient presentation forum produced by the bioluminescent brain, in which all sights, sounds, smells, tastes, bodily sensations, and mental imagery commingle in a continuous, multisensory display. The front half of the presentational sphere is cluttered with colorful visual forms and realistically rendered material objects; here the sense of sight predominates. Occupying the same presentational space—and extending the space fully around the head—is the sonic sphere, within which the placement of sonic objects occurs through a complex binaural analysis of sound waves. In natural settings, this soundscape corroborates and further reifies the presentation of visual forms.

Superimposed upon both of these presentational spheres is another, much more constricted sphere—one that extends only as far as the outstretched limbs—within which the tactile sensations and movements of the physical body are interwoven with the audiovisual realms of perception. Thought, imagination, memory—the elements of an elusive "sixth sense" called *mind*—also freely permeate this presentational space.

As expansive as this projected universe may appear to be—bounded only by the distant illusion of a *sphere of fixed stars*—the wrap-around world hems in the *"self"* in from all sides, its personal space reduced to a small, central void. The hollow, spherical topology of the presentation itself is what *implies* and in large measure *creates* the apparent "you" that dwells at its center. The *self* is an inference drawn from the incurvate structure of the presentation.

The brain is also spheroid in shape and would, I contend, if it could, expand into a perfect sphere of more generous proportions—its grapefruit size and tortuous convolutions being a necessary evolutionary compromise between the unyielding full term human fetal skull and a more constricted bipedal birth canal. In addition to its spheroid form, an extensive and persuasive cluster of features point to the brain's being specifically designed as an organ of *presentation*: the crystalline clarity of the cerebral interstitial and cerebrospinal fluids that bathe the brain inside and out; the transparency of the *pia mater*

THE GLOBE THEATER

(the superthin "protective" casing that surrounds the cerebrum); the vast fields of hair-thin, cortical columns that crowd the entire 6 1/2 square-foot outer surface area of the cerebral cortex in tight, parallel formation, like the multicolored silken threads of a fine Persian carpet; and most tellingly of all, the lavish energy expenditure that results in the "bright" firings of these *presentation neurons*. Weighing in at approximately 1400 grams, a mere 2% of an average human's total body weight, the brain uses up to 30% of the body's nutritional calories to produce its 10-25 watts of "brain power."

The "brain-as-computer" metaphor has reigned so long in the neuroscience community that a host of wearisome prejudices have now become firmly entrenched. "A ghost in the machine" is the witty retort to anyone so naïve as to suggest that the mind might comprise a substance separate from that of the dense, material brain. Then there is that hackneyed, old infinite regress: "If someone is in there looking at *my* brain, then who is in *his* brain looking at *him?*" Well, my new laptop arrived this morning by FedEx, and I'll tell you what—with central processing units (CPUs) powerful enough to perform CAT scans now standard issue on today's notebook computers, my consumer lust was much more focused upon the vibrant, 1920 by 1200 pixel Xbrite™ LCD display and whether or not the soundcard would meet the specs of a high-end multichannel audio system. The real trouble with the "brain-as-computer" metaphor is that it is universally understood to mean the "brain-as-CPU." I have no problem with the metaphor if it includes all the *peripherals*.

This is what has made the mind/brain problem such a bear to unravel. It is as if the pixels of your computer screen were all in cahoots with one another, collectively performing the very computations that orchestrate their panoramic, communal extravaganza. Cerebral cortex neurons are the ultimate multitaskers, simultaneously functioning with equal competence as binary computer circuits, as pixellating elements, and as components of the intricate analogue oscillators and resonators that help generate the sensorium display. Not to mention their ongoing assignment in research and development, where they take on the task of rewiring their own circuitry. There is no clear distinction in the cerebral cortex between processing substations and presentation platforms—it is all one "single-minded" collective enterprise: to create the most compelling model possible—that is to say, a fully integrated, multisensory, and hopelessly *self-centered* view of the surrounding world.

At the very back of the brain is the occipital lobe—the main visual area—which is comprised of a set of *retinotopic* maps, by means of which the external view of the world is accurately translated from the *retinae* at the backs of the eyes onto the primary visual cortex and from there, onward to a set of auxiliary visual cortices. The auditory cortex in the temporal lobes boasts a set of *tonotopic* maps that chart out the sounds of reality's complex "musical score." The sensory and motor *homunculi* accurately map all of the physical sensations and muscular movements of the body onto a pair of "little men" that straddle the tops of the two hemispheres of the brain. Maps, maps, everywhere *maps!* Neuroanatomists have been absorbed for nearly a century in their meticulous task of *mapping* the brain in ever greater detail. Perhaps it is time to abandon this pale, utilitarian term "map" for these brain areas, considering the sublime, transmundane activity occurring here. They are much more accurately conceptualized as the *high-definition plasma viewscreens, surround sound speaker systems, and teams of personal masseurs*, all provided by corporate headquarters for your enjoyment and edification, in the lofty seclusion of your private penthouse suite.

It sounds terribly unsophisticated, anthropomorphizing the brain in this way. The truth of the matter is that the brain's chief function is anthropomorphizing *us!* We are but a nebulous quantum field of electromagnetic *awareness*, brought into existence by the collective firings of neurons. We must be formed and sculpted by specific cortical and subcortical processes into the one particular spatiotemporal configuration expressly "designed" by evolution to seem natural and right to us—one that allows us to perform our day to day corporeal duties untroubled by fruitless existential anxiety. That is to say: a fully articulated *self* embraced by a vivid, lifelike, multisensory world.

"Neo-Cartesian Nondualism"—to coin a somewhat cumbersome phrase—would be the most accurate philosophical expression for this particular brand of mind-brain integration. Mind and brain, though distinct in substance, are definitely not two *independent* entities. Neither are they one and the same. Human consciousness is, in every detail, a construct of evolution, which—like all such constructs—owes its existence to the stepwise unfolding of purely biological processes, pruned by the cruel hand of natural selection. Nevertheless, this grandest achievement of evolution was made possible only because of one curious and irreducible fact of nature: that *bare awareness* ("self-observation") exists in the universe, as an inherent and necessary attribute of the collapsible quantum field. The *photon* becomes the mysterious weaver of the seen and the unseen worlds.

This is the Neo-Cartesian Theater—*Psyche's Palace*—a grand, old edifice with lustrous marble walls, a wide proscenium stage, a full orchestra, and one solitary plush seat in the royal box for Her Highness. If the cerebral cortex is indeed generating this presentational sphere of light and sound and sensation, it stands to reason that the theatrical production team would have already snatched up every spare square centimeter of brain surface area to assist in the display; for this entire worldview should, I contend, map out *directly* upon the convoluted surfaces of the cerebral cortex. But there's the rub. Nothing is where it ought to be in the brain. Although the visual world always projects in *front* of us, the visual cortices are located in the occipital lobe, at the very *rear* of the cerebral cortex. Not only that, but the entire *left* side of the body is presented to and controlled by the *right* hemisphere of the brain, just as the *right* side is presented to and controlled by the *left*. Indeed *every* plausible subjective landmark in the brain is likewise inverted and transposed, back to front and left to right.

Therefore, I conjecture that the *self* must actually ride the brain *backward*, for it engages the widescreen world ahead upon an incurvate visual cortex "screen" located at the very back of the skull. The corporeal sense of *self* begins where the visual realm ends, and the brain conforms to this subjective percept with its transition from the occipital (visual cortex) to the parietal lobe, where visual and somatosensory information is integrated. A *peripersonal map* charts the space occupiable by our own bodies. Its sphere extends only as far as our limbs' reach—and is responsible for, among other things, the visceral sense of "intrusion" when someone enters our *personal space* uninvited. Sensations and movements of our arms, trunk, and legs appear more seamlessly interwoven with the rest of the visual world within the space of our peripheral vision. This multimodal integration of sense fields is definitely parietal lobe activity, and from our reversed perspective, centered in the brain, the left and right parietal lobes would complete the forward half of the sensorium sphere—

evidence that the parietal lobes themselves generate the integrated somatovisual presentation we imprecisely refer to as "peripheral vision."

You can try this experiment for yourself. As your wiggling fingers advance from the far periphery toward the center of focus, does there not appear to be a clear and abrupt boundary where the subjective experience of a fully integrated *somatovisual* sense field gives way to a more purely *visual* perception of your hands as distinct and somewhat autonomous objects? You may notice a disturbing visceral sensation (perhaps, as my friend Julie Minton suggests, a preconscious memory of our original "separation anxiety") occurring at the precise moment when your own hands pass from the oceanic parietal sense field to become, in that instant, discrete visual objects, now clearly and demonstrably *separate* from the *self*.

Passing from the occipital to the parietal lobe, we find ourselves now at the parietal lobe's forewordmost edge; and there, just this side of the central sulcus, smack dab in the middle of the brain, is the sensory homunculus—the corporeal anchor that centers our physical being within the sensorium. On either side of the head, just beneath the temples, one contralateral half of the inordinately large face of the sensory homunculus is permanently planted—bolt upright, well *behind* (from this reversed perspective) and presumably "facing" the visual cortex.

The entire soundscape is reversed in a similar manner, with the auditory cortex for the right ear being located in the left temporal lobe and the auditory cortex for the left ear in the right, like a pair of dorm room speakers switched around on the window ledges, aimed to blast the quad with a prankster reveille. This complete reversal of subjective orientation—forced, presumably, by evolutionary happenstance (the awkward, counterpole positioning of the visual cortex in some preconscious ancestral life form)—neatly accounts for the contralateral organization of the brain, a most conspicuous anomaly for which no other even remotely plausible explanation has ever been offered.

Just as the earthly globe, at any given moment, is illuminated from one side only, so the Globe Theatre of the mind has its dark and secret spaces where the house lights never rise. Our mental percepts (the shadowy thought-forms, amorphous ideas, vague memories, and vacillating intentions) are generated from neuronal firing patterns in the *prefrontal* lobe, which is physically located well in front of the homunculus face. Nevertheless, these mental percepts project to consciousness as *interior* events and sensations, to all appearances originating from within a nebulous "mental realm" subjectively located *behind* the face, *within* the head.

The prefrontal lobe is the headquarters of the executive function; the steady expansion of this brain area marks the evolutionary progression of the hominid lineage, with the rate of growth of the prefrontal cortex accelerating most rapidly in the species Homo sapiens. The prefrontal lobe is the brain's prime real estate—clearly earmarked to accommodate the ever-expanding bureaucracy of the "unitary executive." The runaway development in this particular brain region may have more to do with the restrictive zoning laws encountered in the parietal and occipital lobes. The whole "back forty," so to speak, is already fully developed—and completely devoted to the uninterrupted presentation of the front half of the sensorium sphere.

In Plato's Allegory of the Cave, Socrates describes the plight of mortals in eerily similar terms. He depicts a group of men and women who are chained to the ground, facing the back of a cave. All that they can see are the shadows cast by firelight upon the cave wall of a parade of passersby behind them. They hear the echoes of the others' voices as if the voices were emanating from the shadows themselves. With no other sensory evidence to contradict their limited view, the mortals cannot be dissuaded from their firm belief that they are directly experiencing the real world.

How Psyche came to dwell, upturned and inverted, in her palace of illusion is a story that evolution wrote, sure enough, but in disappearing ink upon the flimsiest of parchment—the soft tissues of the brains of our prehistoric ancestors. The few scattered fossil skulls that have been unearthed—the crumbling ruins of her once magnificent castles—can only hint at the drama that unfolded within them.

The second century North African mystic, Apuleius, delivers a more poetic rendition of the story of the evolution of consciousness in the next chapter, "The Myth of Cupid and Psyche." Science may do well to heed the voices of myth and symbol, for myth and symbol bubble up from the ancient minds of ancient brains and have something of this corporeal truth lingering just beneath their surfaces.

A SUMMARY OF CHAPTER TWO

The chapters of this book present a series of progressive scans of the same complex object—sensorium consciousness in the bioluminescent brain. Each pass is intended to bring new features into focus, while deepening the understanding of material previously introduced. Here, the spotlight falls upon the spherical form of the sensorium presentation, revealing it to be a direct manifestation of the spheroid form of the cerebral cortex itself.

A catalogue of apparent inconsistencies in the topology of the brain has kept its straightforward presentational function well-hidden from scientific discovery. The very same cluster of anomalous features—when viewed 180° in reverse—constellate into a single, clear, and inescapable picture. Consciousness rides the brain backward! Like the prisoners in Plato's cave who can view only the shadows of reality cast upon the back wall and who can hear only echoes, we have no way of knowing from inside of our own sensorium that the soul's orientation is likewise reversed from that of the physical body.

The parietal lobe is theorized to function as a full-service, multisensory integration cortex that melds the presentation of our projected body image with other sonic and visual objects into a single, oceanic somatoaudiovisual space. The prefrontal lobe, on the other hand, occupies the "dark half" of the sensorium sphere; as the nerve center of the hypervigilant executive, it does not specialize in the presentation of reality per se, but uses its best "intelligence estimates" to generate its own set of hypothetical scenarios and contingency plans.

The cerebral cortex is also referred to as the iso-cortex (the "same-cortex") because it has a nearly identical structure over the full six and a half square feet of its surface area. It is not this fact alone but a preponderance of circumstantial evidence that leads me to make this overly bold hypothesis: that most, if not all, of the surface of the cerebral cortex is structured for neurobioluminescent presentation and continuously engages aspects of consciousness in unimaginably diverse modalities of sensory and symbolic display.

CHAPTER THREE

THE MYTH OF CUPID AND PSYCHE

The universe is an irreducible tautology. That is to say, it *is* exactly what it *is*. It unfolds as it will, completely indifferent to what we choose to think or say about it. Yet human beings, desperate for meaning, have always thought and said a great deal about it anyway. Much of human history and culture has been devoted to the contentious group project of developing and promoting various models of the universe and our place in it—with charts and diagrams, pictures and words. We tell ourselves all sorts of stories.

Some stories we call science. These are often the most robust stories, because they are interwoven with logical strands that support one another in a sturdy scaffolding of prediction, experimentation, and verification. With its tools and technologies, theories and hard data, science has allowed mankind to reach high into the heavens, back in time to the dawn of creation, and deep into the unseen world of elementary particles and quantum fields.

Other stories are ethical teachings, either handed down from celestial sources or derived from universal truths. From them we learn how to behave in ways that help us move gracefully through the world and strengthen our bonds of community and friendship.

Yet not all stories are true, of course. Some of the most powerful stories that guide the lives of many poor souls are nothing more than a pack of lies, scare tactics, and political spin—horror stories and apocalyptic visions of doom. They too serve their purpose, and they serve it all too well.

My purpose—in writing this book—is to tell a different story. As it turns out, it is a rather long and winding tale, but one that has given me a lot of pleasure and comfort and a sense of connection to the world I live in and the people I love. Whether it is true or not, I cannot rightly say. The story begins with the age-old question—the riddle of the Sphinx. "What goes on four legs in the morning, on two legs at noon, and on three legs in the evening?" Man, of course, who crawls, then walks, then hobbles from birth to death, his three or fourscore years likened to a single day.

What does it mean to be human? With our fragile bodies suspended between the vastness of frozen space and a whirling ball of magma, finding sanctuary within this thin slip of biosphere—our every act is a gesture of faith. It is all so very unlikely. We come here from God knows where, we live a modest or magnificent life of this and that, and before we know it, we are returning to the mystery, vanishing as if in a puff of smoke. And all along the way, from the earliest childhood fairy tale, through the doctoral thesis, and up to and including the administration of last rites, it is all stories within stories within stories.

The purpose of *this* story, and there is one, is to place something plausible between the realm of the body and the realm of the spirit—a credible conduit between the corporeal and the incorporeal—because if these two are separated by an impenetrable barrier in life, then we are forever split in our loyalties. If the body is here and the spirit is there, but the body will eventually die here, then shouldn't I be focusing my attention *there*? But the things and people I love are *here*. And what if there really is no *there* there?

On the other hand, if one can develop the sense that there truly is an organic enmeshment between the body and the spirit and that it can be understood and experienced immediately, here and now, in innumerable ways, then life and death become more fluid companions as well.

THE MYTH OF CUPID AND PSYCHE

In the empty chasm between the here and the hereafter lies a great deal of anxiety for most people, it seems. Without a compelling theory, a modern myth to bridge the gap, many people adopt a cocky, feigned indifference, turning away from spirituality because it seems embarrassingly simpleminded—too primitive and superstitious in this age of cell phones and particle accelerators.

Not that we understand cell phones and particle accelerators, mind you, but we have faith that someone does, and this faith in the modern world gives us comfort. Somebody must know where this train is headed, right? All that is left to do is to stuff the anxiety down, conceal it in a pile of consumer goods, and turn our attention back toward the trivial.

But let's not go that route. There is something really wonderful just down this other path. Let's take a look.

Some years ago I happened upon the myth of Cupid and Psyche by Apuleius, the second century North African mystic and philosopher. Cupid is the Roman name for Eros, the Greek god of erotic love. Psyche is the soul. This is the story of how they became entwined in marriage. These ancient myths often come from a place of deep intuition in the human unconscious, and the play of archetypal images within them can be highly instructive. You may be wondering where all of this is heading. Just sit back and relax your mind—soon it will all be clear.

—There once lived a king and queen who had three beautiful daughters, the youngest of whom was possessed of a beauty beyond anything that the poverty of language could describe. Her beauty was of such renown that people were drawn in great crowds from neighboring lands to pay her a degree of reverence that was rightly due to the goddess Venus alone. The temples of Venus were left vacant, as men converted in their devotion to this virgin girl. When she walked by, the crowds would sing her praises and toss flowers at her feet.

Such veneration toward a mere mortal was an unseemly offense to the true goddess Venus, who exclaimed in indignation, "Am I to be vanquished by a mortal girl? I will give her good reason to regret flaunting her unlawful beauty."

She called to her mischievous winged son Cupid and pointed out Psyche to him, saying, "My dear son, you must punish that obstinate beauty. Revenge your mother as severely as she herself has been injured. Cause a passion to breed in the heart of that arrogant girl for some unworthy, despicable being, that she may feel a humiliation as abject as her current triumph is glorious."

Preparing to obey his mother's command, Cupid entered Venus's garden to fill his two amber vials—one with sweet water and the other with bitter—from the two fountains that flowed there. He tied the vials to his quiver and flew off to Psyche's bedchamber, where she lay sleeping. He poured a few drops of the bitter water over her lips, regretfully, for her beauty had almost moved him to pity, and then lightly touched her side with the tip of his arrow. At this touch she suddenly awoke, startling the invisible Cupid, who in his confusion pricked himself with his own arrow. Wishing to undo his mischief, Cupid poured out the waters of joy over the curls of her silken hair.

Venus's curse had its desired effect upon poor Psyche. Though she was greatly admired for her beauty, no man in the kingdom would ask for her hand in marriage. Her parents consulted the oracle of Apollo only to receive this dreadful reply: "Your daughter is destined to marry a hideous monster whom neither gods nor men can resist. She must hasten to the top of the mountain to meet him, for he eagerly awaits his bride."

With her fate confirmed by the oracle, Psyche submitted and ascended the mountain to the summit. From there she was carried down by the gentle Zephyr into a flowery meadow where she soon discovered in a nearby grove of trees a magnificent palace built by Cupid. Drawn toward the building by admiration and wonderment, she entered and found it filled with a great variety of objects of delight. There was an enormous vaulted roof supported by pillars of gold, and the walls were adorned with exquisite sculptures and immense landscape paintings depicting herds of animals and rural hunting scenes, all delightful to behold. There were other rooms in the palace, filled with treasures and beautiful works of art.

An invisible voice interrupted her reverie, stating: "Sovereign lady, everything you see here is yours. We are your servants and will obey all of your commands with the greatest diligence and industry." Psyche sat down in an alcove where a table magically appeared, spread with all manner of sumptuous delicacies and glasses of nectareous

THE MYTH OF CUPID AND PSYCHE

wine. A chorus of invisible singers, accompanied by a lute, filled the air with glorious harmonies.

She never saw her husband, for he came to her only at night, always fleeing before sunrise. His voice was full of passion, and soon she felt a similar passion for him growing within her breast. He would not allow her to see him, though she often begged him to let her do so. "Why would you wish to behold me? Do you have any cause to doubt my love? If you did see me, perhaps you would fear me or adore me as a god. But I only wish to have you love me as an equal."

Psyche's concerns were somewhat assuaged by his words, and for a while she felt reasonably happy. After a time, though, she grew homesick for her parents and sisters and began to feel as if her palace were nothing more than a gilded prison cell. A visit with her sisters was arranged, during which Psyche began to doubt her own description of her husband, that he was a beautiful youth who spent his days hunting in the mountains, and feared that he might be, as her sisters warned, a ferocious monster, intent upon doing her grievous harm.

Psyche tried not to think about this, but after her sisters were gone, their words began to prey upon her mind. Her irresistible curiosity prompted her to devise a means to discover for herself just who her husband actually was. She hid a lamp and a sharp knife beside her bed, and after he had fallen asleep she arose to uncover the lamp. She beheld not a gruesome monster, but a beautiful god with golden curls, rosy cheeks and snow white wings upon his back.

As she leaned over him with the lamp to have a closer look at his face, a drop of burning oil fell and scalded his shoulder. He awoke, startled, fixing his eyes dead upon Psyche. Without a word he spread his wings and flew out through the window. Psyche tried to follow him, but fell from the window to the ground below. Cupid turned back for a moment in flight and said, "O foolish Psyche. Is this how you reward me for my love? I disobeyed my mother in marrying you, and now will you imagine me to be a monster and cut off my head in my sleep? Go back to your sisters, whose counsel you prefer to my own. I will not punish you other than to leave you forever. Love cannot abide with suspicion."

After many adventures, trials and tribulations (to which we will return in a later episode), the lovers finally reconcile, winning the consent of Venus and the blessing of Jupiter, king of the gods, who presents Psyche with a goblet of ambrosia, saying: "Take this and drink, Psyche, and be immortal. Cupid shall never break free from this knot of love in which he is tied. Your nuptials shall be perpetual."

Thus did Psyche become united with Cupid, and in the course of time they gave birth to a daughter whose name was Pleasure.

I haven't included these long passages from this ancient love story simply to entertain you. The myth of Cupid and Psyche happens to be a perfect synthesis of nearly all of the themes we are going to be considering in this book. What a classical myth can do far better than a scientific theory is to penetrate deeply into the recesses of your *own* psyche, preparing the soil for a subsequent, more sober investigation of the material with the logical mind. Reading this tale of the two young lovers, you were introduced to the concepts of body and mind as personifications, while in a more generous mind state—curious, perhaps, to discover how these two may have become entwined—and less likely, therefore, immediately to arouse the chopping and slashing activity of the analytical mind.

The challenge now is to unwrap the meaning. The following paragraphs will give a suggestion of some of the themes that will be fleshed out in succeeding chapters. First, to our two protagonists.

Psyche, in the original Greek, means breath, from *psychein*, to breathe. It is similar to the Latin *spiritus*, a breathing, from which we get the words spirit and inspiration. Psyche is the personification of the soul. She descends from the heights of the mountaintop carried by Zephyr, the god of the west wind. Her story is meant to teach us something about our own soul and its descent into incarnation.

The name Cupid may conjure up images of the rosy-cheeked baby cherub with a toy bow and arrow on a valentine, but that emasculated picture comes from a later, more prudish era in history. Recall that Cupid is identical with the Eros of Greek mythology—a powerful, strapping young man, the god of erotic love and sexuality.

The most important clue is given immediately in the story, and that is that Psyche is extraordinarily *beautiful*. Psyche is the mind, but her chief attribute is not intelligence. Her quarrel is not with Minerva, Goddess of Wisdom, but with Venus, Goddess of Beauty. In this modern, scientific age, we immediately associate the mind with thought, logic, intelligence, analysis. The myth brings us back to an earlier time, when the mind was more closely associated with aesthetics. Here we have an allusion to the concept of the brain as a neural resonator, more akin in function to a finely crafted musical instrument than to a computational device.

Cupid's palace represents the sensorium, the theater of the mind produced within the brain, a palace Psyche enters willingly, enticed by the beauty it presents to all of her senses. That this palace was built by Cupid reminds us of his role in the great wheel of cyclical existence. Attraction, desire, lust, and erotic love set in motion the chain of events that leads to sexual reproduction—the biological mechanism that links one generation to the next—allowing for natural selection and the further evolution of the body, the brain and the mind.

In the palace (incarnated in the perceptual sense fields of the body-mind), Psyche is the lady of the house, with invisible servants (sense organs and other corporeal functions) providing for her every need. Occasionally, she feels trapped by her circumstances, a prisoner in her own palace. This hints at a darker aspect of the soul's dilemma in its incarnation, which we will investigate more fully in chapters eight through ten.

Psyche's quandary is that she does not know whether her husband is truly the beautiful youth to whom she is attracted, or the hideous monster that fills her with repulsion. Here we have the perfect encapsulation of what in Buddhism are known as the Three Poisons: greed (or desire), hate (or aversion), and delusion (or ignorance). They are seen to be the primal characteristics of mind prior to enlightenment, the cause of the endless striving that perpetuates the wheel of birth and death. Cupid poured upon the sleeping Psyche both sweet and bitter waters (of desire and aversion) from the two amber vials he had filled at the fountains of Venus's garden. The darkness of the bedchamber represents the ignorance that keeps the mind of Psyche vacillating between fear and longing—the perfect metaphorical image for the predicament of the embodied soul.

When decoding this myth, it is important to remember that Venus is a goddess, and when a goddess of her stature decrees something, it happens. The Oracle of Apollo is also unerring in its prophecies. Venus's curse and the oracle's prediction have therefore, we may presume, come to pass. Psyche *has* married a hideous, unworthy, despicable—but nonetheless irresistible—monster. He is also the beautiful youth who inspires love and desire in her. This is a perfect depiction of the soul's relationship with the physical body. The body provides a palace of pleasure for Psyche, but one that can in the blink of an eye be transformed into a merciless prison of pain. The body keeps this treasure of awareness enthralled, in both senses of the word: enchanted and enslaved.

Psyche's insatiable curiosity represents our own desire to understand the mysterious relationship between the soul and the body. Her courage to approach the quandary with the lamp of clear seeing and the knife of daring discernment finally triumphs, overcoming her sentimental attachment to the familiar palace/prison of ignorance.

Whether prompted finally by anxiety, curiosity, or a fervent desire to know the truth, Psyche finally discovers that an open-eyed, fully conscious—*awakened*—relationship with the body and its passions is possible to attain. She learns that transcendence is not a movement up and away, but downward, fully integrated into the flesh, in loving partnership with the physical body. And the offspring of this union is Pleasure.

And this, I deeply feel, is our assignment as well.

THE MYTH OF CUPID AND PSYCHE

A SUMMARY OF CHAPTER THREE

The myth of Cupid and Psyche is introduced. These characters will reappear throughout the book to illustrate aspects of the relationship between the body and the mind in much more visceral, human terms. In the myth, Cupid, the god of carnal love, builds a palace for the mortal Psyche, the personification of the mind of beauty and light. Cupid's palace represents the body, in particular the brain and the neuronal structures that support and constrain the sensorium presentation. This "living palace" provides Psyche with every sensuous delight but can also be experienced as an oppressive prison that she is never permitted to leave.

The vacillations in Psyche's mood highlight a curious facet of this new theoretical perspective—that the primary function of even the most primitive sentient animal mind is fundamentally "aesthetic." Desire (liking) propels movement toward an object; aversion (disliking) propels movement away from an object. These are the two most basic and universal functions of animal life. The ability to move toward and to move away from is, after all, what distinguishes mobile animals from their firmly rooted plant cousins.

Cupid is revealed to be both a loving husband and a monstrous tyrant who keeps Psyche in the dark as to his true nature. The relationship between Cupid and Psyche, the body and the mind, is strained. Psyche's curiosity to discover who her husband actually is represents our own thirst to understand what it means to be a spiritual essence housed within a physical body.

CHAPTER FOUR

NEUROBIOLUMINESCENCE
A BRIEF INTRODUCTION TO THE THEORY OF NEUROBIOLUMINESCENCE IN THE EVOLUTION OF SENSORIUM CONSCIOUSNESS

This chapter is intended to serve as a very thin guidebook for the voyage ahead. These few pages will certainly not suffice as a complete explanation of anything; I can only hope that they will whet your appetite for what is to follow. The full set of ideas and insights that constitutes this theory cannot easily be contained within a single frame of reference. The theoretical model freely cross-mates elements of philosophy, science, spirituality, and the arts to form a strange gestalt. For the object of this theory, *consciousness*, is itself a peculiar and elusive beast, only discernible in its native habitat if viewed simultaneously from several different perspectives.

The theory entails precisely one "metaphysical" postulate: that aspects of the quantum field itself—in this case, the minute quantum electromagnetic field effects produced by the firings of presentation neurons within the brain—are the foundational source of a basic *proto-awareness* from which our own highly articulated conscious awareness is formed. The wave-particles—the photons, electrons, and positrons that make up the quantum electromagnetic field—exhibit capricious, probabilistic, nonmechanistic behaviors that appear to many quantum physicists to correlate with fundamental aspects of consciousness. The deeper metaphysical question of how this proto-consciousness came to dwell within the quantum field is not the subject of the present theory, however. What concerns us here are the mechanisms by means of which the evolving biophysical organism has been able to utilize this latent proto-consciousness for its own ends—how the basic raw material was molded by evolution into the first sentient beings. This theory posits that the ultimate biological function of the sentient brain is to organize and amplify the minute, proto-conscious electromagnetic field effects of its firing neurons into the very scene that manifests each moment before our eyes, that fills our ears, and through which we appear to walk—a three-dimensionally superimposed, multisensory presentation that effectively re-creates an entirely credible view of the surrounding world within the dome of the skull. This sensorium is a miraculous simultaneity—experienced by the awareness inherent within the very quantum electromagnetic field that comprises the presentation itself!

All of the sights, sounds, smells, tastes, thought forms, and feelings that constitute the hypothesized sensorium are represented by a hodgepodge of highly specialized and diversified quantum electromagnetic field presentation structures, the vast majority of which render their artful compositions directly upon the convoluted surfaces of the cerebral cortex. The sense modality most clearly correlated with electromagnetism is of course *vision*, which translates the electromagnetic patterns of visible light that strike the retina onto a set of cortical maps whose neurons' electrochemical firings coalesce to form an electromagnetic *field* that precisely mimics the visual scene. This is what makes vision the easiest mode of sensorium presentation to conceptualize. The eight specialized visual cortices located at the back of the brain are entirely analogous to fully embodied Technicolor movie screens, although the *self*/world-creating moving pictures of the visual cortex are much more refined, being further heightened in their realism by a vast amount of cortical and subcortical image processing.

It may be somewhat less obvious how the lightbody experiences sounds, smells, tastes, textures and emotions *electromagnetically*—or as aspects of some perceptible quality or dimension of light. Vision is certainly the most powerful and compelling (and most mysterious) of the sense modalities. Consciousness itself can be rightly understood to be an *involution* or *recapitulation* of the selfsame electromagnetic field from which all visual data is drawn. Until we get a better "feel" for how the other five senses might be displayed within the brain's complex electromagnetic field—and how they would be experienced by an indwelling consciousness—other aspects of the sensorium are best understood as hypothetical analogues to vision.

This new theory challenges a cornerstone assumption of neurobiology—that all neurons are engaged in one of only a few fundamental activities: first, sensation—the initial gathering of sense data by specialized neurons within the sense organs; second, communication—shuttling information from one part of the brain to another or sending and receiving signals to and from the rest of the body; or third, some form of computational analysis—processing, storing, and retrieving information in large groups of associated neurons that behave like complex neurobiological computer circuitry. This theory posits a fourth and fifth function for particular groups of specialized neurons. The fourth function involves an extensive class of *presentation neurons*, organized into a wide variety of cortical and subcortical "maps" that operate quite literally as presentation screens. These neurons are responsible for generating the electromagnetic field displays of all modalities of sense perception. These sensory presentations commingle to form the illusory, circumambient sensorium that surrounds—and in so doing defines—an introversive "self-center."

The fifth category of neurons are those arranged in feedback loops (as described in the first chapter of this book) that act as *neuronal resonators* of sensory perceptions in cortical areas devoted to aesthetic evaluation and certain aspects of presentation. These neuronal resonators form highly integrated structures composed of more basic *neuronal oscillators*, arrangements of simple rings of fixed numbers of neurons completing circuits of sequential firings at a given rate, allowing them to sustain particular frequencies. Intricate combinations of these basic components have evolved into elaborate mechanisms for the analysis of musical intervals, harmonic structures, visual patterns and proportions, mathematical relationships, and geometric forms. Collectively, these neurons form a multisensory *neural resonance chamber*, which, like a violin's sound box, provides temporal continuity for consciousness and a rational basis for the aesthetic evaluation of sense data.

This theory defies all standard neurobiological models of brain function. It rests upon the assumption that in the case of presentation neurons, the firing of the neuron is an end in itself—that the firing is perceived by consciousness in the same manner that one might perceive and distinguish a red or green pixel on a television screen. The specific neuronal structures that could easily produce this sensorium presentation are already well known to exist within the visual cortex. It has always seemed odd to me that this simple inference has never been carried all the way through to its logical conclusion—but if researchers have had no reason to look at these neuronal structures as the pixellating presentation elements of sensorium consciousness, it is understandable how they could have overlooked them. None of the structures or processes involved in sensorium presentation are extraordinarily difficult to imagine, nor do they appear to require any new or exotic neuronal components. Their evolutionary development, as we shall see, can also be readily accounted for.

Once again, the single metaphysical postulate undergirding the theory is this: the source of awareness within the brain is the quantum electromagnetic field produced by the firing of its neurons. Electromagnetism being a cumbersome and inelegant word, more associated in the imagination with hydroelectric turbines and household gadgets, I replace it with the more spiritually evocative term *Light*. In common usage, the word "light" refers only to *visible light*, that portion of the electromagnetic spectrum perceived by the human eye, the full rainbow of colors from extreme red, with a wavelength of approximately 7700 Angstroms—through orange, yellow, green, and blue—to extreme violet with a wavelength of approximately 4000 Å (400 billionths of a meter or 400 nanometers). This actually represents only a small fraction of the full electromagnetic spectrum, which includes gamma rays with wavelengths of less than 0.1 Å; x-rays; ultraviolet, visible, and infrared light; microwaves; and radio waves, with the lowest frequency radio waves having wavelengths of 100,000 km or more.

The brain is constantly generating from 10 to 25 watts of electrical power, converting up to 30% of the body's available nutritional calories into these individually minute, but cumulatively staggering electrochemical voltages. The quantum field within the brain is rightly viewed primarily as an *electric* field. However, the flurry of tiny movements of charged sodium and potassium ions through the electric field during neuronal firings in turn generates its own minute *magnetic* field effects—and all such interactions between electric and magnetic fields are mediated by real or virtual *photons.*

Light, in physics, is a somewhat ambiguous term that can refer specifically to electromagnetic radiation within the range of 4000 to 7700 Å (violet to red), which is perceptible to the unaided human eye, or more generally to electromagnetic radiation of any wavelength. The physicist and the neurobiologist would agree that the brain generates an electromagnetic field—that is not in dispute—but both would be equally perplexed by my persistent use of the term *Light* for that field. A brain surgeon is not required to wear protective goggles when he peeks inside your cranium; the cerebral cortex is not pulsing with a bright neon glow. Both would view the brain's electromagnetic field as mere evidence of entropy, analogous to the heat given off by a motor, which does no useful work. I use the term *Light* intentionally—to *cast light* back upon the field itself as a legitimate field of study—by evoking the word's ancient associations with spirit and consciousness.

At this early stage of our investigation, it is not essential to know the precise energetic composition of the brain's quantum field—it seems sufficient for our purposes to recognize that it is a vibrant, supple, malleable field of *energy*, distinct from—yet clearly produced by—the physical brain. The more immediate question is how this field accomplishes the presentation, how it re-creates the actual shapes and textures of reality through the patterned firings of its neuronal substrate. If we can think of the brain's quantum field as being analogous to ceramic modeling clay, what is of greatest interest to us now is the range of *sculptural* possibilities inherent in such a medium; whether its actual substance—beneath the shiny glaze—turns out in the end to be earthenware, porcelain, or terracotta is not our immediate concern.

Topologically, each hemisphere of the cerebral cortex can be conceptualized as a crumpled two-dimensional surface, displaying a wide variety of sensory *cortical maps,* which I contend function as the brightly lit, constantly refreshed "presentation screens" of all sense modalities. The microscopically fine-grained structures of the brain's electromagnetic field vary point-to-point and moment to moment across the surfaces of these cortical presentation screens as their constituent neurons fire and fade. A cascade of rapidly opening and closing voltage-gated sodium ion channels in the neuronal cell wall—

the biomechanical device that causes a neuron to fire—increases the intensity of the electromagnetic field at that particular point in time and space. This means, at the quantum physical level, that there is at that moment and location an increase in the quantum probability of the four basic electromagnetic field effects: the creation, annihilation, attraction and repulsion of photons, electrons, and positrons. On the "spiritual" plane, this neuronal firing represents an infinitesimal and localized "brightening" of a tiny constituent of conscious awareness.

Awareness is therefore taken to be a fundamental property of the quantum electromagnetic field, associated with the photons and charged wave-particles produced in such abundance by the brain. However, until funneled through the neurobiological mechanism, this basic awareness bears absolutely no resemblance to what we experience as fully elaborated sensorium consciousness. When a wave-particle is created during a neuronal firing, it issues forth as a quantum of awareness from the plenum void. The rudimentary nature of this awareness cannot be overemphasized. It is as insignificant as the tiniest speck of amber pigment on the cheek of the Mona Lisa. Nevertheless, without this tiny speck of paint and billions more like it—and without this quantum of awareness and billions more like *it*—neither the artwork nor the artist could ever have been created.

It would be nice if this were the full extent of the complications, if we could at this point simply posit a more or less permanent conscious viewer within the brain whom we might imagine to be seated calmly in his cerebral armchair, watching these presentation screens and listening to the piped-in music. But the universe is much more creative than that. Consciousness, it turns out, is not "housed" within the brain; it flows *through* the brain as if emerging from a cornucopia. Consciousness is always arising fresh in each moment from the mysterious plenum void. It is continuously reformulated into this fleeting and illusory presentation of a *self* and a world surround. There is nothing permanent about our form of consciousness except for the mechanism by which we are drawn into being—a biophysical apparatus as ancient as embodied consciousness itself.

Throughout this book I have tried to maintain essential moorings to those venerable spiritual traditions that have earnestly sought to reveal the mysterious machinery of consciousness. The Buddhist traditions are preeminent among them—their founder, in my opinion, being the first truly rational spiritual scientist. The Buddhist worldview is large and wise and compassionate enough to allow us the space to process the disquieting news about who we actually are and how we got ourselves into this predicament. Most importantly, it offers guidance and practical techniques for liberating ourselves into a new and fully engaged life of happiness and ease.

Advanced practitioners of certain Buddhist introspective meditation techniques who claim to have experienced the normally preconscious phase of arising consciousness attribute three fundamental and irreducible qualities to this newly emerging "quantum" of awareness: *ignorance*, *desire*, and *aversion*. Awareness always arises in an initial state of blinding bewilderment at the moment of its emergence. It has no idea who or what it is—it is only aware of its own becoming. This is the source of its "affinity" for light and for further becoming and its "disinclination" to return to darkness and annihilation. (Please recall that this is no more than a whirlwind introductory tour of some very complex ideas that will be more carefully introduced in chapter six.)

I contend that the body has learned to capitalize on these three fundamental attributes of the light of awareness, evolving a mechanism that utilizes specific neuronal firing patterns to guide each new iteration of awareness into being. We will look at one theoretical model in particular—with roots in the ancient Buddhist *Abhidharma* tradition—in which consciousness emerges like the flickering frames of a motion picture in a continual succession of moments of creation. Each conscious "entity" is carried along a pathway that retraces the stages of both the evolutionary and embryonic development of the brain. Originating as a flash of bare awareness within the reptilian brainstem, it then passes through the old mammalian limbic structures, until it is fully elaborated within the neocortex, where the final illusion of a wholly realized body and self-concept is experienced, appearing within a sensorium—a world surround apparently populated by other objects and persons.

The mechanism I will propose directly parallels the theory of *microgenesis* (which was just outlined above), a groundbreaking model of consciousness first introduced by Jason W. Brown, M.D. in his 1991 book (retitled, 2002), *The Self-Embodying Mind: Process, Brain Dynamics and the Conscious Present*. His theory is based largely upon meticulous observations made in his neurological clinic of the specific cognitive deficits displayed by patients suffering from a variety of brain injuries and diseases. This conceptual model provided me with one of the final pieces of the puzzle, and although his theory flies in the face of standard neurobiological connectionist models, it nonetheless corroborates many of the intuitions of the perennial philosophy and has become an important element of my own thesis.

What is new in my own theory of *photogenesis*—though absolutely in accord with Brown's theory of *microgenesis*—is the understanding that the consciousness that arises in this manner is actually *Light* itself, an elaborate electromagnetic field fluctuation produced by the collective firings of many billions of neurons, forming a "lightbody" that emerges and reemerges moment after moment from a single point of origin in the brainstem, expanding in complexity, fleshing itself out, as it were, as it retraces the evolutionary and developmental pathways of the brain, all within a fraction of a second. The unimaginable intricacy of this expanding lightbody is what actually produces the sensorium, creating within itself its own immediate experience of the moment. It is a developing awareness, composed entirely of *Light*, aware of its own shape, texture, color, and sound, which is progressively molded by specific neuronal structures and processing devices within the brain until it becomes the perfected model of a *self* and world surround that we experience as waking consciousness—and which I refer to as the sensorium.

The sights, sounds, smells, tastes, bodily sensations, memories, and dreams—as well as the I-thought that you yourself are presently experiencing—are composed entirely of *Light*. This *Light* is the medium that links matter and consciousness into a functional nonduality. Not separated into observer and observed, *Light* is simultaneously both the entirety of the continually shifting display of the firing neurons responsible for the creation of the sensorium and the totality of conscious awareness that experiences the illusory display.

In Brown's theoretical model, neuronal activity must be viewed as fundamentally informational, communicational, and computational in nature (although the sense of all of these terms and the diagram of his "circuit board" are entirely at odds with existing connectionist models, which his theory effectively displaces). Yet it remains unclear how the *experience of consciousness* could ever enter any such closed system of interneuronal activity. The physical pathways that this information must travel along are clear enough—chains of neurons passing signals along their axons, firing across synapses to dendrites of other neurons, whose axons are linked to other dendrites of neurons farther down the line. Presumably, in Brown's microgenetic model—through some as-yet-unexplained *gestalt* process—consciousness would simply "emerge" as a result of the hypercomplexity of these interconnections. What better "informational environment" to *emerge from* than the electromagnetic field itself, whose full spectrum light, since the beginning of time, has radiated throughout the universe, pinpointing stars, bending around gravitational lenses, ricocheting off planets, and scattering through our atmosphere, revealing in intimate detail the colors, forms, and textures of objects near at hand and far, far away?

Brown's theory, which will be examined in more detail in chapters five and sixteen, describes exceptionally well the developmental sequence that each thought-moment undergoes as it emerges from a pure, undifferentiated "wakefulness" in the brainstem. It first enters dream consciousness, where it mingles with sense perceptions, developing into an initial object awareness, finally becoming a fully realized analytic perception, actualizing the separation into a *self* and a world. Brown is perhaps too wise to speculate—in that book, at least—upon precisely *what it is* he believes is emerging; but fools like me rush in!

In my conception, it is the *Light* itself that travels along these evolutionary and developmental pathways. Although an electromagnetic voltage does course along the neuronal axon, this is most likely not of sufficient electromagnetic "substance" to sustain the contours of an expanding lightbody. The development of this conscious entity, according to the present theory of neurobioluminescence and photogenesis, is not constrained to movement along these interneuronal connections, although it certainly can utilize them. It is the neuronal cell body itself, in particular the *axon hillock* that is the brightest pixellating element for all developmental phases of sensorium presentation. This is the most excitable part of the neuron containing the highest density of sodium channels in its cell membrane. It is the location of the greatest concentration of electric charge during neuronal firing and hence the site of the largest fluctuations in electromagnetic field strength. It is where the hundred millivolt amplitude *action potential* originates before it propagates along the axon to the presynaptic terminals that connect to the dendrites of other neurons.

The theory of photogenesis will receive a much more thorough treatment in chapter fifteen, which is entitled "The Stream of Consciousness." Until then, I hope that this brief synopsis will suffice. I propose that the brain has evolved a mechanism for shepherding proto-awareness into full sensorium consciousness from an initial burst of *Light* within the brainstem. The emerging lightbody is guided along an ever expanding, brilliantly lit canal through the length of the brainstem as if being expelled from a womb (a most apt comparison, as it turns out). One may also visualize this movement as being analogous to peristalsis in the esophagus, the wavelike muscular contractions that allow food to be guided from the mouth to the stomach while the muscles of the throat themselves remain in place.

The brainstem, I surmise, has neuronal structures that fire in a predictable sequence, producing an illusion of moving lights, not unlike an elaborate Las Vegas casino marquee. The pace of "development" of the emerging entity may be fine-tuned by evolution, just as the speed of the apparent movement of lights on the marquee may be adjusted by its designer. The illusory movement is produced by the sequential firings of neuronal cell bodies in close proximity.

In principle, there may be no direct axonal-dendritic connections whatsoever between subsequent neurons in the series. In this sense, adjacent neurons are precisely analogous to adjacent light bulbs on a casino marquee, which are not directly connected in series, permitting their blinking rates to be controlled by more sophisticated electrical timing mechanisms located elsewhere. A pulse of *Light* produced by the initial "pacemaker" neurons deep in the brainstem is succeeded by a unidirectional series of firings. This presents an irresistible, brightly illuminated "path of escape" to the newly emerging unit of conscious awareness. A dark, unappealing void surrounds the moving entity on all sides, and an advancing darkness follows close behind it, like the sequence of esophageal constrictions that guide food down the throat.

The present theory posits that within the cerebral cortex the firing of sensorium neurons creates a luminous pixellation that is directly experienced by awareness. The visual, auditory, and somatosensory worlds of sight, sound, and bodily sensation come into being by means of this unmediated communication of *Light*. The cytoarchitecture—the cellular structures—of the brain reflects the primacy of presentation, as evidenced by the abundance of "presentation screens" (which have traditionally been referred to as cortical and subcortical "maps") that are found throughout the brain. There have always existed strong evolutionary and developmental pressures to improve the quality of the presentation—constantly expanding the neuronal repertoire of subtle electromagnetic effects to give the display a more lifelike look and feel. Periodic evolutionary upgrades in the biocomputer's hardware and software packages (the proliferation of neuronal cell subtypes and improvements in their organization) are what give the body its controlling advantage over the soul. The brain has the ability to capture the attention of awareness, since the final presentation of the sensorium is entirely determined by the physical "wiring" of the neurons and the strategic information being passed between them.

It is hard to imagine that lowly electromagnetism, the force that runs your vacuum cleaner, garbage disposal unit, and hair dryer may actually be responsible for the arising of human consciousness. By an accident of technological history, its mundane applications were thoroughly capitalized upon by industrialists and entrepreneurs long before its mysteries were revealed through the esoteric equations of quantum mechanics and string theory. Perhaps we would do well to return to the mindset of a Benjamin Franklin, Galvani, Volta, or Faraday, when the forces of electricity and magnetism were just beginning to be scientifically conceptualized and were still cause for reverent awe. Once in the hands of Thomas Edison, electromagnetism was destined to become simply the workhorse of industry, stripped of its mystery and any shred of proper dignity.

For a theory of consciousness to be viable there must be a plausible accounting for its evolution upon this planet, as either an "emergent" phenomenon arising from completely nonconscious sources, as a development of latent consciousness within the physical or biological substrate, or as a last resort, as a prefabricated "drop in" from some parallel spiritual dimension. In this theory, sensorium consciousness is not prefigured in any way but rather proceeds in a straightforward, rational progression of evolutionary, biophysical advancements, each a logical outgrowth of prior nonconscious, computational processes in earlier, preconscious animals.

These animals were also completely ignorant of and indifferent to any of the "spiritual" aspects of electromagnetism. Like Edison in his laboratory, they evolved only practical uses for electricity, first in the form of simple neurons that connected their archaic sensory apparatus (eyes, ears, antennae, skin) directly to their organs of motility (fins, legs, wings). Primitive brains developed as intermediary processing stations, charged simply with the task of correlating sensation and movement. These brains evolved in complexity for some time as the purely functional, computational hardware of what were essentially robotic automatons completely devoid of sensorium consciousness. These mobile units may or may not have been graced with any form of awareness or selfhood—the present theoretical model assumes that they were not. The story of how presensorium, cybernetic (information processing) brains developed sensorium consciousness requires a great deal of background preparation. For now, suffice it to say that the transitional mechanism proposed in this theory will be airtight and unassailable—assuming, that is, that the one metaphysical postulate mentioned above passes your personal muster.

We will also grapple with the unconventional view that all animals, not just those of the human variety, are possessed of their own unique complex of aesthetic sensibilities. Far from being a recent refinement, aesthetics represents the very crux of sensorium consciousness. Our bodies go to the considerable trouble of providing sustenance and lodging for an indwelling conscious entity in order to be able to exploit that entity's innate capacity to distinguish what it likes from what it dislikes. Some configurations of its electromagnetic form—certain transient neurobiochemical milieux, certain arrangements of objects in the sensorium sphere—it finds preferable to others. Desire and aversion, wrapped up in a tight bundle of delusion that is mistaken for selfhood—these are the three primal forces that maintain the basic structure and functionality of the "sentient being." Gaining a fuller understanding of the role of desire and aversion in the evolution of aesthetics—and the elaborate emotional states that have evolved to support these subtle movements of mind—awaits a more extensive treatment of the complexities of brain chemistry in chapter nine.

In the human being, this aesthetic sense has completely transcended its utilitarian, evolutionary purpose, and now we may bask in the resplendence of art, music, theater, and dance, enjoying cleaver wordplay while gathered together at sumptuous feasts, taking leisurely walks through heavenly scented gardens, and opening our hearts and minds to the hidden dimensions of reality and imagination. The theory of neural resonance provides a sound theoretical framework to account for this notorius proclivity of the human being to indulge in such luxurious, opulent excess. Evidently the brain is not, as we may have once suspected, a cold and calculating computer—not by any means. In form and function it would be more fittingly compared to a warm and vibrant musical instrument.

At some point, though, we may find ourselves in the grip of an existential angst—as the Buddha did prior to his awakening—realizing that beneath this manifest beauty there is a basic level of unsatisfactoriness that threatens to undermine our attempts to find abiding happiness within the sensual pleasures of this fleeting world. Interspersed throughout this book are suggestions, guidance, and words of ancient wisdom intended to bring us to a mature accommodation of our lot—and a joyful expression of our potential—as human beings. I refer to this practice that aims to penetrate the illusions foisted upon us by our own body-mind (without introducing new sets of "spiritualized" illusory constructs) by the playful, pseudo-Germanic term *"Realmeditazion"* (meditation upon unvarnished reality). The method can be useful in gently, progressively stripping away our sentimental attachments to the illusory aspects of the "self-construct," while simultaneously avoiding any possibly lethal flirtation with a bleak and nihilistic *existentialism*.

Once again, it is an actual physiological function of the brain that provides the guidance for how to address and overcome the existential dread of being cut off from the "real world" within one's "prison skull." The newly emerging iterations of *self* arising moment after moment from the brainstem are not, as is postulated in some ancient Buddhist texts and mirrored in Jason Brown's theory of microgenesis, completely detached from one another in discrete and isolated moments of Now. The neural resonance chamber is the sanctuary within which all of these "separate" beings merge and blend. Through the practice of "self-welcoming," we learn how to work with the body-mind that we actually have, to become the gentle, peaceful community of mind that welcomes into the world, in every moment, each new incarnation of ourselves. In so doing, we create the conditions that expand our capacity for peaceful coexistence with all beings in all realms.

Yes. Happiness is possible. Enlightenment is inevitable. You'll see!

PSYCHE'S PALACE

A SUMMARY OF CHAPTER FOUR

This theoretical model of consciousness rests upon a single "metaphysical" postulate: that the quantum electromagnetic field is the source of a basic proto-awareness from which our own highly articulated conscious awareness is formed by evolutionary forces.

The sensorium is the quantum electromagnetic field produced by the firings of neurons that mimics the shapes and textures of the forms perceived in the outer world. The quantum field presentation is self-aware—its powerful sense of subjectivity due in large measure to the topology of the presentation—a hollow, spherical surround—from which it infers the self/world dichotomy and its own self-centrality.

The cerebral cortex is, topologically, a crumpled two-dimensional surface, overlaid with a wide variety of sensory cortical maps, which function as brightly lit, constantly refreshed "presentation screens" of the primary sense modalities.

Awareness always arises in an initial state of blinding bewilderment at the moment of its emergence. It has no idea who or what it is—it is only aware of its own becoming. This is the source of its "affinity" for light and for further becoming and its "disinclination" to return to darkness and annihilation.

The Light itself, an elaborate electromagnetic field fluctuation produced by the collective firings of billions of neurons, takes the form of a "lightbody" that emerges and reemerges moment after moment from a single point of origin in the brainstem, expanding in complexity, "fleshing" itself out, as it retraces the evolutionary and developmental pathways of the brain, all within a fraction of a second.

Our bodies go to the considerable trouble of providing lodging for an indwelling conscious entity in order to be able to exploit its innate capacity to distinguish what it likes from what it dislikes. Desire and aversion, wrapped up in a tight bundle of delusion that is mistaken for selfhood—these are the three primal forces that maintain the fundamental structure and functionality of the "sentient being."

CHAPTER FIVE

CONCEPTUAL FRAMES

As the elegant clockwork model of the Newtonian universe has given way to quantum mechanics and superstring theory, the new conceptual models that have arisen to replace it seem ever more arcane, counterintuitive, and paradoxical. Some physicists and cosmologists are looking eastward to Taoism, Buddhism, and Hinduism, hopeful that the commonalities between the Eastern religions and Western physics might provide an ideal philosophical framework for modern science. Early popularizations of the East-West science and spirituality dialogue, such as *The Tao of Physics* by Fritjof Capra or *The Dancing Wu Li Masters* by Gary Zukav, have given us the clear impression that Physics, the currently reigning queen of sciences, is the most obvious choice to be the emissary to the East and to enter into scholarly conversation with their masters of meditation.

Focusing on Buddhism's elaborate cosmologies and its invocations of the mysterious Void, physicists were quick to latch on to apparent, but perhaps ultimately spurious parallels—in equating the Buddhist concept of "emptiness" with the "quantum vacuum," for example, or in comparing "quantum paradoxes" to Zen koans. What is so easy to overlook in such a remote and abstracted analysis of the Dharma is the centrality of suffering—the great quandary of the human condition—at the core of the Buddha's teaching.

Although she is far down the line of succession in the conventional academic hierarchy, I would like to put in my enthusiastic recommendation for Evolutionary Biology, the queen's less glamorous distant cousin, to serve as emissary, particularly in the Buddhist context. I firmly believe that if the Buddha could somehow have been slipped a copy of Charles Darwin's *The Origin of Species*, he would have found within it an ample scientific foundation for his views on sentient existence, the nature of suffering, and the path to liberation.

The Buddha was a scientific thinker, in what most people would consider a prescientific age, when philosophy and science were more closely allied, and both were regarded as contemplative arts. If we were to apply his Law of Dependent Origination from a more rigorous, modern perspective, investigating thoroughly with all of the technological means at our disposal, we would come to precisely the same conclusion that he did: that no form or structure in the universe, be it physical, biological, or conceptual, exists in isolation. Nothing is permanent and independent. All objects, all beings, and all mental formations are interconnected; they exist subject to causes and conditions; and they are all caught up in the same web of basic insecurity. No composite thing can offer an ultimate, abiding refuge, including, most poignantly, the *self*.

Emptiness of self nature, impermanence, and *unsatisfactoriness*, the Three Marks of Existence, which the Buddha maintains are characteristic of all possible phenomena and all composite things, can easily be misunderstood to have been intended as purely theoretical postulations upon the nature of the physical universe. Yet his singular purpose, as always, is to elucidate the illusory nature of the *self*. This contemplation is part of an extended exercise in inductive reasoning, working from the knowable (the physical universe, the sense organs, thoughts, and feelings) toward the more elusive and problematic (the ego, the I-sense, or the *self*). After exhaustive consideration of countless examples of dependent origination in the *physical* world, upon what grounds can one grant one's *self* an exemption from such a basic and universal principle? What you take to be *you* is likewise empty, impermanent, and unsatisfactory.

CONCEPTUAL FRAMES

The Buddha's teachings are intended to awaken the practitioner by exposing the *self* as a fabrication—the *self* that arises in the form of sentient beings, these physical and mental muddles with sense organs and nervous systems in which we are entangled. My question is: what happens to the Buddhadharma, the essential teachings of the Buddha, if, instead of relying on the ancient cosmological, metaphysical, *karmic* framework in which these ideas first arose, we make a modern hypothesis—that what the sentient being is experiencing, fundamentally, is a *biological* entrapment. I am asking you to imagine what Buddhism might have looked like if the Buddha, somehow, in the year 500 B.C.E., had possessed the tools and insights of evolutionary biology to account for the gradual enmeshment of the body and the mind. If one can fully understand the biological basis for the predicament of the sentient being, one will have yet another incredibly powerful conceptual tool to pry one's own mind free from its corporeal restraints.

It may sound as though I am trying to dismiss physics somehow—as if that were possible in a theory that posits *light* as the conduit of the soul. But that mysterious veil will surely be the last to fall. Light is a particle and a wave—and neither and both—a quantum event. Light is both information and the limits of information imposed by the uncertainty principle: the paradox trigger that keeps Schrödinger's cat—hapless victim of the infamous thought experiment—in life-and-death limbo. How some bandwidth of the quantum electromagnetic field of light contrives to function as the carrier wave of the noösphere (the ostensible planetary "mental aether") is a deep mystery, well worthy of scientific and meditative study. It is what the ocean is to the ichthyologist, however, or the atmosphere to the bird watcher—an important adjunct to the central discipline. This encounter of mind and body takes place squarely in the squiggly, wet world of *biology*. Biology is nested in organic chemisty, and organic chemistry is nested in physics. Physics is the odd bird that has no nest so claims the whole universe as its lawful territory. At the end of the day, it is not the soul's familiarity with the equations of physics that allows her to romp in the fields of incarnation, no more than does a laughing baby, splashing about in its bath, need instruction in fluid dynamics to do so.

It is a subject of... let us say... lively debate in Buddhism just what it is that travels from incarnation to incarnation: whether a mindstream, a bundle of karmic tendencies, or something called the *jiva*—a provisional, noneternal soul, distinct from the *atman,* the eternal soul of the Hindu Vedas, which is roundly and vehemently negated *(anatman* being the Sanskrit Buddhist term for *no-self).* Here I must invoke a personal agnosticism. I don't know if I *am* the soul, if I *have* a soul, or to what degree what I think I am or have is a delusion. I am *aware.* An adjective. I don't know about aware*ness,* the noun. I use the word *soul,* in its full fuzziness, simply as a convention.

There are three tracks that lead to the soul's incarnation, and they wind through separate territories of the scientific terrain, each operating within a vastly different timeframe. The first track ushers us through the province of evolutionary biology—a journey of many hundreds of millions of years—during which time we witness the origin of the first sentient beings: from the body's initial tentative encounters with awareness to its eventual capture and containment. Carrying us onward through the full elaboration and refinement of the sensorium, the first track culminates in the creation of these astonishingly complex beings that we now are—or believe ourselves to be.

The second traces the soul's journey through the life cycle of this particular body-mind—the gestation and development, the creative encounter with form, the life-learning and contemplation that have brought each of us into this present moment of mature self-reflection. Here we traverse the realms of embryology, developmental to transpersonal psychology, and cognitive science.

The third track careens wildly, as the moment to moment experience of the soul as *mind,* as mediated by the brain and nervous system awash in the flow of neuropeptides throughout the body. This is the province of neurobiology and its numerous frontier sub-disciplines.

There is a formula first expressed by German biologist and philosopher Ernst Haeckel that he called the "biogenetic law," a now discredited hypothesis in biology that nonetheless survives in the popular imagination, largely because of its curious turn of phrase and its evocative power in linking, or attempting to link, two great disciplines with a single concept. "Ontogeny recapitulates phylogeny" is the expression. The theory failed only because of an overambitious hypothesis: that *all* of the stages in the development of the embryo of a species repeat *all* of the stages in the evolutionary history of that species.

It nonetheless contains more than a kernel of truth and is at least a partial explanation for many of the commonalities that are seen in embryonal development across widely divergent species. It is now generally accepted in biology that species sharing a common ancestor pass through similar stages in their embryonal development. The backbone, for example, is a structure common to all vertebrates, including fish, reptiles, birds, and mammals, and therefore is one of the first structures to form in all vertebrate embryos. The embryos of all land vertebrates, including human embryos, still show gill pouches at one stage of development, evidencing an ancestral link to fish. The newest and most sophisticated feature of the human brain, the cerebrum, develops last in the embryo.

The striking parallels between *ontogenesis* (the development of an individual organism) and *phylogenesis* (the evolutionary history of the species) are best explained by the neo-Darwinian theory of evolution—where new species evolve through a series of small modifications in the genetic program that guides development. Genetic mutations that are expressed *early* in fetal development are more likely to cause catastrophic failure by their disruptive effects on all later stages. Mutations that are expressed after well-tested archaic structures are in place are more likely to prove viable and perhaps advantageous. Therefore, the cutting edge of evolution tends to be a tinkering with the most recent structures—maybe a sleek new profile, a chrome grille and tailfins, but the same old engine and chassis as last year's model.

Jason W. Brown, in *The Self-Embodying Mind: Process, Brain Dynamics and the Conscious Present*, introduces a third parallel to *ontogenesis* and *phylogenesis* with his provocative theory of *microgenesis*. It is a neurologically based theory, but one that does not concur with the conventional models that describe information processing modules interconnected in multidirectional neuronal communication networks. Instead, the overall neural sequence branches out like a tree, traveling unidirectionally from the most archaic structures to the most modern, recapitulating all of the stages in the evolutionary history *and* fetal development of the human brain.

The theory states that global firing patterns lasting only fractions of a second emerge from the depths of the brainstem (the reptilian brain), pass upward through the limbic system (the old mammalian brain) and continue on to the surface of the neocortex (the new mammalian brain), traversing all of the successive layers of brain structures in the same order in which they were laid down during the course of evolution and fetal development. The process is described by Deane Juhan in his introduction to Brown's book as "the rapidly flickering recapitulation of an individual's entire past as the context in which each moment of the now is experienced." The impression of continuity of experience is a deception caused by the continuous replacement of the overlapping representations rising out of the core. "Pacemaker" neurons in the brainstem are hypothesized to exist that would initiate a new sequence approximately every tenth of a second.

This is a dynamic model, in which objects of consciousness begin their journey in a formless matrix of pure wakefulness, devoid of image or object. The impulse elaborates in the brainstem into an initial preobject, which is passed forward to the limbic structures where the object begins to form, still undifferentiated from the *self*, and takes on emotional affect. After several other stages of microgenesis, in which the object is categorized and sculpted by processes of the parietal cortex, it is passed on to the neocortex where the full presentation of the world, its objects, and their relationships to the *self*, have completely matured.

The extent to which sensory input constrains the emerging object varies according to mental, emotional, and environmental factors. Deane Juhan summarizes this point in his introduction to Brown's *The Self-Embodying Mind*:

> *Note that the emerging object is not made up of assembled sensory information, but that sensory information is one of the elements at work in the formation of a specific perception. Overall contexts, feeling tones, and categorical associations also play decisive roles in determining what it is that finally emerges as the current self, object perception, and world surround...*
>
> *This is the stage in which projection plays a powerful role in the emerging perception. Our categories and comparisons from which a type of object is selected are based on past experiences, and as such constitute the contents of our expectations and beliefs about the world. This is why it happens so often that novel occurrences are not even seen, but are merely slid past without notice, reshaped to fit past experience, or simply denied.*

CONCEPTUAL FRAMES

Jason Brown's theory of *microgenesis* is compatible with many of the basic tenets of Buddhism, as articulated in the collection of early Buddhist philosophical and psychological writings known as the *Abhidharma*. It is even more closely linked to the profound *mahamudra* tradition of Tibetan Vajrayana Buddhism. Again, in the words of Deane Juhan:

> *Each of these moments of microgenesis—literally, "little birth," or "now-realization"—is the perceptual emergence of an entire self and a world, a continually resurrected self and world that take the place of the ones immediately prior to it, and which die away as immediately, making room for the next. In this continual self-emergence, the whole of the individual's past is continually deposited into the virtual duration present at the rim of the future that does not yet exist, and each of these moments contains the germ of novelty within the current never-before-experienced now.*

I was at first reluctant to inject what many may regard as "religious scriptures" into the mix, but I suppose I might be a bit naïve in thinking that I have avoided anything scandalous in my presentation so far. Those of you who are still reading have already shown a great deal of patience with some rather unconventional ideas. Yet I feel that I am still only laying the groundwork, even at this point in chapter five. I cannot imagine how it would be possible to conceptualize the thesis in its entirety without first referencing some core Buddhist ideas.

To insist that Buddhism is simply a philosophy and not a religion seems a bit contrived. It *is* a philosophy, certainly, but one that does not limit its scope to the everyday facts of consensus reality. It weaves in realizations and insights only attainable after many years of intensive meditation practice. Reincarnation, for example, in most Buddhist cultures, is taken almost as a self-evident fact of life, based upon the reputedly authentic reports of those who have experienced it in full consciousness. The 14th Dalai Lama, so it is claimed, is also the 13th and the 12th and the 11th and the 10th...

My own faith in Buddhism derives from the clarity of its philosophy and the efficacy of its meditative techniques, first and foremost.

After examining its conceptual framework from all sides and in all situations, one begins to appreciate its universality and explanatory power. From this experience a sense of admiration naturally arises for the dedicated practitioners both past and present who have received and cultivated these insights. In some this develops into a religious impulse toward the Buddha—the progenitor of these liberating practices—in others not.

I have spoken so much of the soul in this story so far that you may be wondering how this squares with the central Buddhist doctrine of *no-self*. I have used the terms soul and Psyche because they are beautiful, evocative words, with a positive emotional association. When I use the word soul, I don't know if I mean anything substantially different from ego or *I-sense*. And the ego, from a Buddhist perspective, is the culprit responsible for all of the mischief. Buddhism may be the religion of kindness, but it is ruthless in ferreting out ignorance and would certainly view this sentimental attachment to poor, dear Psyche as very sloppy thinking.

What Buddhism denies, and it does so forthrightly and without compunction, is the notion of a *separately existing self*. The ego or I-sense is simply an elaborate ruse. It is an impermanent structure, built from causes and conditions and held together largely by erroneous beliefs. You are not the ego, regardless of how much you may believe yourself to be. You are something much better, much freer. You are, in essence, pure awareness, the *wakefulness* that in some traditions is called *Buddha nature*—unborn, primordial and causeless—the foundation of mind from which the ego emerges through an elaborate process, a kind of comedy (or tragedy, more aptly) of errors.

The most lucid account of the development of ego that I have ever come across appears in the first chapter of *Glimpses of Abhidharma*, by Chögyam Trungpa, meditation master, scholar, and founder of Naropa University. We will begin the next chapter with a "glimpse" of these pivotal Buddhist teachings.

CONCEPTUAL FRAMES

A SUMMARY OF CHAPTER FIVE

The most profound philosophical question of all—how bare awareness itself came to exist within the substratum of the universe—may be well beyond the scope of human understanding. In this chapter, the view is put forth that everything shy of this Eternal Mystery may actually lie well within the realm of the understandable.

The specific formulation of conscious awareness that we, as humans, are endowed with has been forged by hundreds of millions of years of Darwinian evolutionary pressures. We cannot find in this world examples of consciousness existing in the abstract; the human mind has evolved into its current form through natural processes deeply embedded in the wet world of biology. Sensorium consciousness is evolution's practical application of a potentiality already inherent in the understructure of the physical and energetic universe.

The links between Buddhist thought and evolutionary biology are first broached in this chapter. Buddhism's goal is to understand the true nature of reality and in so doing to alleviate the root causes of unhappiness. The Buddha's Three Marks of Existence—suffering, impermanence, and no-self—are all irrefutable biophysical realities, enunciated expressly to thwart our futile attempts to find enduring happiness through the usual channels: by satisfying all desires and avoiding all aversive situations. This sobering troika makes short work of our infantile dreams of Heaven (a place of Eternal Happiness for Me). We can gain fresh insights into the Dharma through the scientific investigation of consciousness and its pivotal role in evolution.

Jason Brown's theory of microgenesis is highlighted in this chapter because it offers a key insight into a possible mechanism for linking together three seemingly unrelated fundamental processes essential to the evolution of conscious, animate life. He shows how the entire evolutionary history of our species is not only paralleled in the stages of growth of a fetus during the course of its development, but that this recapitulation also happens continuously within the brain—moment after moment—propelling ever fresh iterations of consciousness through the same evolutionary and developmental sequence.

CHAPTER SIX

EGOGENESIS

Chögyam Trungpa, in his book, *Glimpses of Abhidharma*, gives us a lively and colorful introduction to these classic texts, the earliest expositions of Buddhist psychology. This particular set of teachings focuses on the process Trungpa calls "the origin of the neurotic mind." Lest you think he is speaking about some crazy "others"—know that from the point of view of the *Abhidharma*, all minds, prior to awakening, are understood to be foundationally neurotic. This may seem, on the face of it, to be a purely theoretical construct—a hypothesis arrived at by logical inference. In the Buddhist understanding, however, the whole of the *Abhidharma* represents a distillation of countless meticulous observations of internal mental processes made while in deep states of meditative absorption.

What is unknown—and perhaps unknowable—is just what level of reality is being experienced here. As you read these descriptions, bear in mind that there are at least three distinct possibilities. First, it may be that the clarity of insight attainable within these deeper planes of consciousness allows the practitioner to perceive directly some fundamental structure of the universe itself—an aspect of the plenum-void, the "full-emptiness" of the quantum vacuum from which "virtual particles" emerge and to which they return.

The second possibility is the one proposed by the school of *yogacara*, a form of Mahayana Buddhism, whose complex cosmology can also be summed up neatly in the maxim: "All three worlds [of desire, of form, and of formlessness] are mind only." This would be consistent with Trungpa's *mahamudra* perspective and is clearly the one he is advancing here. There are two formulations of the "mind only" position: the stronger—that the entire universe *is* a projection of the mind—and the weaker—that all that can be *known* of the universe is the mind.

The third possibility is that these experiences are actually perceptions of processes occurring deep within the brain itself, as consciousness retreats in meditation from the fully formulated neocortex projections of the waking sensorium, beneath even the archetypal dream space of meditative visualization—tracing with awareness even further down into the ancient recesses of the brain, prior to the formation of objects, prior even to the formulation of the *self*.

If we call the process *egogenesis*, we may more readily see the parallels to *microgenesis*, *ontogenesis* and *phylogenesis*. As you contemplate the following paragraphs, imagine that these are the "three worlds" that Psyche is entering and that the same forces propel her into existence in each of these three disparate time frames: the moment, the lifetime, and the infinite stream of incarnations.

The *Abhidharma* is an analysis of the process of ego formation. The *ego* is that which forms in an attempt to reconcile desires and conflicts relating to the outside world. It is the mediator that bears the burden of the uncertainty as to whether these desires and conflicts reflect the actual condition of a real world, or whether they are internal and unreal—paranoid projections of the mind. The root of all psychological problems is in this perpetual endeavor to determine what is real and what is unreal.

The *Abhidharma* examines the formation of the ego from the point of view of *egolessness*. Egolessness is a way of describing the condition of the mind after it has accepted the realization of the actual nature of the world—that it is both transitory and transparent. Directly seeing the interdependence of the "inside" and the "outside," the ego's projections are recognized to be insubstantial, and the ego itself becomes progressively more transparent and transitory—a harmless dissipation.

The ego is comprised of eight modes of consciousness, according to the *Abhidharma*. There is a consciousness for each of the five senses and one for the thinking mind, which is regarded as a sixth sense. The seventh consciousness is a unifying structure, as odd as that sounds, considering that it is basically a cloud of ignorance and confusion that permeates the first six sense consciousnesses. It is within this cloud of ignorance that the fundamental error, the separation of *self* and *other*, occurs. The seventh consciousness is the source of the erroneous discrimination between the "inside" and the "outside." The seventh consciousness concludes that the eighth consciousness is the *self*, which it clings to, believing it to be a permanent refuge. It distinguishes the *self* from the *world*, the contents of the first six consciousnesses.

The eighth consciousness is the unconscious, common ground that allows the other seven to operate as a functional whole, by providing the illusion of separation from the even more basic *ground of existence*, which contains everything, including both *samsara* (the cycle of birth and death) and *nirvana* (liberation). The eighth consciousness is distinct from this basic ground; at the level of the eighth consciousness confusion has already begun, and it is within and due to that confusion that the other seven consciousnesses operate.

Trungpa states:

> *There is an evolutionary process which starts from this unconscious ground, the eighth consciousness. The cloudy consciousness arises from that and then the six sense consciousnesses. Even the six senses evolve in a certain order according to the level of experiential intensity of each of them. The most intense level is attained with sight which develops last.*

Out of the basic ground of existence arise the energies that are the source of all relative situations, as well as the ego that mediates them. Described as "sparks of duality, intensity and sharpness, flashes of wisdom and knowledge"—everything emerges from this basic ground.

The eighth consciousness comes into being when the flashes of energy arising out of this basic ground cause a bewilderment—and a blinding panic. It all happens too fast; and the bewilderment, fearful of returning to its source, latches on to the developing present moment and clings to it. In this first instant of creation, ignorance, aversion, and desire arise almost simultaneously, propelling this proto-egoic entity through the inevitable course of its development.

From this basic level of *form*, the development of the ego is traced through the other four *skandhas* (the "aggregates" that constitute the personality) of *feeling, perception, intellect,* and *consciousness.* This is as far as we will take the Buddhist analysis of the development of ego, except for the following comments Trungpa makes to clarify the distinction between the *Abhidharma* and *mahamudra* teachings. This distinction, I believe, is what accounts for the wide discrepancy of programs and practices for liberation that have been developed in the various schools of Buddhism. The sexual practices of a "left-handed Tantrist," for instance, contrast sharply with the celibacy of the Thai monk, yet both are pursuing the path to liberation.

> *The difference between the Abhidharma and basic Sutra teachings on ignorance and the more direct and daring mahamudra teaching is that the Sutra and Abhidharma teaching relates to ignorance as a one-way process—bewilderment and grasping and the six sense consciousnesses develop and ignorance takes over. But in the Vajrayana teaching, ignorance is seen not only from the angle of the development of ego, but also as containing the potential for wisdom.*
>
> *The wisdom of dealing with situations as they are, and that is what wisdom is, contains tremendous precision that could not come from anywhere else but the physical situations of sight, smell, feelings, touchable objects, and sounds. The earthy situation of actual things as they are is the source of wisdom. You can become completely one with smell, with sight, with sound, and your knowledge about them ceases to exist; your knowledge becomes wisdom. There is nothing to know about things as an external educational process. You become completely one with them; complete absorption takes place with sounds,*

> *smells, sights, and so on. This approach is at the core of the mandala principle of the Vajrayana teaching.*

As pertinent as this aspect of the Abhidharma teaching will prove to be, vis-à-vis the theory of neurobioluminescence in the evolution of sensorium consciousness, there are two other foundational Buddhist teachings that I believe need to be looked at with a much more skeptical, modern eye. The first is the ancient Buddhist cosmological claim that all manifestations of material reality—the mountains, the oceans, the stars, the entire universe—arise within the mind itself and that the physical body and brain as well are therefore understood to be manifestations or projections of the mind. The second concept is the inexplicable and elusive, yet all-powerful and universal causal force known as *karma*.

Just as the clear and liberative message of Jesus has in many peoples' minds been compromised by its association with a scientifically untenable six-day creation cosmology—an ancient tribal myth and ethical teaching story that was mistaken for literal gospel—so do these prescientific Buddhist concepts threaten to undermine a modern student's faith in what would otherwise be a perfectly sound and rational Dharma. Fortunately, the teachings of the Buddhadharma, as per the instructions of its founder, are not to be taken on faith alone, but are to be examined carefully and tested within one's own experience—before determining whether or not to adopt them as a true guide for living.

From its inception, Buddhism has had a paradoxical relationship with the mind, which is seen both as the source of all suffering and as the vehicle for liberation. Beginning around the year 150 C.E. and continuing up until the fourth century, the development of a number of important Mahayana texts, including the *Sandhinirmochana,* the *Lankavatara,* and the *Avatamsaka Sutras,* resulted in the creation of the *chittamatra* ("mind-only") or *vijnanamatrata* ("consciousness-only") cosmologies, which expanded and formalized tendencies already present in early Buddhist theory. The term "mind-only" may at first appear innocuous enough. It may even be an attractive contemplation to think that there is no real world "out there"—that it is all an illusory projection of the mind. But the ramifications of such a worldview can actually be quite pernicious, completely anesthetizing the compassionate heart and inducing a sense of dreamy unreality and disconnection from worldly affairs. Although many other Buddhist concepts—of compassion, loving-kindness, generosity, and patience—help counteract these ill effects, there is no reason to cling to such dubious theoretical constructs if they are not essential to the teaching.

It is my contention that the "mind-only" cosmology is one of the two most serious errors or, if you will, *overstatements* in Buddhist philosophy. The other is the slippery and imprecise notion of *karma*. Karma is posited without proof as the Grand Unified Theory of Everything. It reduces all of the complex moving parts of the universe—the multitudinous physical, chemical, biological, psychological, sociological and cosmological dynamics of cause and effect, in all their chaotic profusion—in other words, the forces that actually make the world go 'round—into one vast field of precise cosmic justice. Karma is granted both an ultimate, enigmatic unknowability on the one hand and an absolute moral overlordship on the other. This formulation appears to me to be wildly speculative. Karma is seen as the operative agent behind the Law of Dependent Origination—also known as the Twelvefold Chain of Causation. Although this teaching presents a precise and accurate description of a local phenomenon—the chain of events that perpetuates the cycle of existence through individual reincarnation—its applicability to and purported unification of all scales of time and space may be (...how to put this?) somewhat exaggerated.

It is well past time, I contend, to question these outmoded Buddhist concepts if they are interfering with a clear perception of the *suchness* that science, philosophy, and religion all claim to seek to know. We need to recognize that Darwinism poses just as serious a challenge to these venerable Buddhist ideological constructs as it does to the creationist cosmologies of fundamentalist Christianity.

There is a slippery slope in the karmic, mind-only philosophy that appears to have prompted the following *well-intentioned*, yet utterly unskillful sermonette I overheard not long ago in a Palo Alto café. I paraphrase broadly for theatrical effect, but the gist was: "The cancer from which you are dying is entirely a product of your mind, and if you could only purify your thoughts sufficiently, then your body—which is also a product of your mind—would naturally produce the spontaneous remission you so desperately hope for. And you will have to do something about the desperation of your hope, mind you, for that in itself is a mental block that creates this imaginary 'disease' from which you are choosing to suffer."

Thomas Byrom's rendering of the first lines of the *Dhammapada: The Sayings of the Buddha* gives some indication of where such perturbations in the mind of this Buddhist practitioner may have had their unintended genesis. Byrom entitles this first section *Choices*:

> *We are what we think.*
> *All that we are arises with our thoughts.*
> *With our thoughts we make the world.*
> *Speak or act with an impure mind*
> *And trouble will follow you*
> *As the wheel follows the ox that draws the cart.*
>
> *We are what we think.*
> *All that we are arises with our thoughts.*
> *With our thoughts we make the world.*
> *Speak or act with a pure mind*
> *And happiness will follow you*
> *As your shadow, unshakable.*

Of course the Buddha is, as usual, "spot on" correct. I take his words to mean that my habits of thought create my own experience of the world. When he says, "With our thoughts we make the world," I feel no compulsion to take him literally. This is, after all, poetry.

A too literal reading can sanction behaviors that are both injurious and unjust. Blaming the victim for the entirety of his predicament—for his poverty, for his ill health, or for the crimes committed against him—is the elephant in the corner (that someone has tried to conceal with a small, safron doily) of the spacious and well-appointed living rooms of those who hold to a "mind only" cosmology. This widely sanctioned Buddhist view generally remains politely unspoken when ministering to the less fortunate, for it would obviously be incredibly bad form to point out the cosmic justice of the situation to someone if he is deemed incapable of making sufficient strides in purifying his mind to see an actual difference in the outcome—but nonetheless, the poor creature is encouraged to do what he can.

Additionally, if the world is considered to be in its entirety a mental projection of a deluded mind, then the complexity and specificity of our physical forms, these actual bodies that we supposedly create from our karma—with their organs, skeletal structures, circulating fluids, and countless intricate biological processes—are seen as living proof that the mind itself that produced them must be equally convoluted, labyrinthine, highly structured, solid, and intractable in its *delusions*.

Here is a typical instance of the invocation of karma as an explanation for the arising of the world and the body's physical form. It is taken from "Part Four: The Eighth Consciousness" of *Verses Delineating the Eight Consciousnesses* by *Tripitaka* Master Hsuan-Tsang of the Tang Dynasty, translated and explained by Ronald Epstein:

> *The eighth consciousness contains seeds, karmic potentials created by previous karmic activities. The seeds ripen and become actual dharmas* ["phenomena" or "constituent factors"] *as they are "perfumed" by the karmic activity of the first seven consciousnesses. The image here is built on an analogy with sesame seeds, which take on the fragrance of the sesame plant's flowers or of any fragrance with which they come into contact... Alaya means "storehouse." Because it is a "storehouse" of seeds, storehouse consciousness (alaya-vijnana) is one of the names by which the eighth consciousness is known... It undergoes perfuming and contains the seeds both of the body with its organs and of the material world. The body with its perceptual organs and the entire physical world also arise from seeds contained in the eighth consciousness.*

The doctrine of karma did not originate with the Buddha. It is a foundational concept in the sacred Hindu texts called the Vedas, which had already been in existence for between 700 and 1000 years. Fundamentally, karma is a neutral system of cause and effect. It is understood to be an inherent property of the space-time continuum. It implies reincarnation as a means to explain the fortunes and misfortunes of one's current life circumstances. In Buddhism—as distinct from Hinduism, where all karma is equally treacherous—karma can be described as either *good* (liberative) or *bad* (binding). It could in fact be argued that the Buddha's teaching was an important reform of the doctrine of karma. From the Vedic standpoint, all karma is binding karma, and ascetic withdrawal from all activity is considered the only true path to liberation. In the Buddhist view, the path includes the performance of meritorious acts of kindness and generosity toward one's fellow beings, from which one gains meritorious karma.

When asked what he would do if it could be scientifically proven beyond a shadow of a doubt that reincarnation was impossible, the current reincarnation of the Dalai Lama is reported to have said that he would immediately stop believing in it. But how, he inquired, would you go about proving that? This is an example of the genuine open-mindedness and spirit of inquiry that has always attracted me to Buddhism.

I am not proposing anything quite as radical as the abolition of karma. It is still a fine "catchall of causality" for whatever universal patterns and processes modern science cannot account for. In my experience, karma is as active as ever in the moment to moment unfolding of my daily life, often hoisting me on my own petard, administering the swiftest possible justice when I transgress. "What goes around comes around" and "You reap what you sow" are good guidelines to live by, regardless of whether there is an actual "bean-counter in the sky" meting out absolute cosmic justice. I think that this fixation upon a purely mechanistic, legalistic, judgmental brand of karma plays into our natural penchant for wanting inescapable punishment for others' transgressions and guaranteed rewards for our own good behavior. Perhaps there are other, more complex reasons why things work out the way they do. Maybe chance plays more of a role in our lives and in the universe than we are comfortable considering.

What I am suggesting is that evolutionary biology already has a proven mechanism for storing the instructions for creating all kinds of bodily structures—and mental tendencies as well: selfish behaviors, violent predispositions, lewd and lascivious drives. The genes themselves that encode the basic blueprint for the creation of this bodymind are also quite capable of storing "habit energies," fear response patterns, delusions, aggressive tendencies—all sorts of challenges to confront the indwelling awareness. The legacy of each individual's struggle for survival remains as an indelible imprint upon the ancient structures of the DNA molecule itself, which acts as an ancillary "storehouse" for much of what has been attributed in Buddhist philosophy to karma.

Additionally, there are two vast treasure troves: the first of hard-won intelligence, resourcefulness, and creativity and the second of emotional sensitivity, affection, and social conscience. Both of these we have inherited from our mammalian, hominid, and recent human ancestors. They have been carried forward to the present day in the genes and, more recently, in cultural *memes* (an analogous term used to describe the mechanism by means of which successful ideas spread through human cultures). They can be thanked as much as any concept of "good karma" for providing us with the basic tools needed to cultivate and ultimately meld these treasure troves of Wisdom and Compassion into the long-sought unity of awakened mind.

The reason why Buddhism should be able to survive the challenge from evolutionary biology is that at the core of the teaching is the understanding that all sentient beings are involved in an "evolutionary" process. There is the sense of an overall steady advancement, as the mindstream develops and learns, as it makes better and better choices in its behaviors—though disastrous setbacks are possible at every turn. Buddhist philosophy, at a loss for an alternative mechanism, relied upon the concept of karma to account for this continual metamorphosis of the mindstream—in its steady advancement through a progression of ever more auspicious rebirths—as it wends its way onward toward its inevitable liberation.

Recall that it was only in 1859—with the publication of Charles Darwin's *The Origin of Species*—that the world was introduced to the radical notion that all of the life forms we see around us (including the ones through whose eyes we see!) are the handiwork of evolution by natural selection. It was also only in the mid-19th century that Gregor Mendel's experimentation with artificial plant hybridization opened up the era of speculation on the mechanisms of heredity that led to the eventual discovery in 1953 by James Watson and Francis Crick of the double helix structure of deoxyribonucleic acid, or DNA.

The ancient Buddhist logic was impeccable, but without the fossil record to consider, and with no conceivable biophysical mechanism to account for the inheritance of physical, psychological, and emotional predispositions, the venerable philosophy of karma provided, for more than two thousand years, a seemingly solid footing for further conjecture.

I am not proposing some scheme that allows us to shuck our responsibility in those countless instances where we truly are reaping the consequences of our actions. Yet much could be learned from making a more nuanced examination of the wide spectrum of evolutionary, developmental, and sociological factors, to determine which aspects of our moral character may rightly be seen as the inheritance of karma and which aspects have a more mundane explanation. Many of the troublesome structures of the body-mind have been inherited from hundreds of millions of years of evolutionary struggle in real life mortal combat. It's time we begin to let the *body* carry *some* of the burden of its own complicity. Although I admit I'm a coconspirator, I'm no longer willing to take *all* the heat.

What I have discovered to be true for myself, is that everything that I find liberative and uplifting about Buddhism is still firmly in place after switching from the ancient to the modern cosmology. The understanding of the physical mechanisms that sustain the universe may change radically, while all of the well-established meditation techniques, the heart-opening practices, and the possibilities for freedom and happiness remain completely intact.

This is especially so in regard to the Four Noble Truths, the core of the teaching, the clearest possible description of the insight that actually constituted the awakening of the Buddha's mind. We will be examining this key to liberation much later in the book, in chapter eighteen. For now, let's just take some time to enjoy our newfound respectability. Difficulties will continue to arise in our lives, as always, but they should not be seen as incontrovertible proof of a basic unworthiness, of a lack of accomplishment, or of a poor attitude. It's much subtler than that. And once removed from that hook that was not of our doing, we may find that we are much more willing to take on the task of cleaning up what actually *is* our own mess.

EGOGENESIS

A SUMMARY OF CHAPTER SIX

This chapter may appeal more to the diehard Buddhist scholar or to those interested in the dialogue between Western and Eastern philosophical positions on the nature of mind. Notwithstanding, the relevance of these time-honored concepts to the present theory is clear. A brief overview of the traditional Abhidharma analysis of the origin of consciousness is given first, for comparison with Jason Brown's microgenesis model.

Objections are raised to the cittamatra ("Mind Only") view that permeates much of Mahayana Buddhist thought. Although our minds can only experience a secondary representation of the world carried in through the senses, this by no means indicates that the underlying reality of the physical universe should be brought into question. It is true that the mind presents us with a version of reality that is in many ways biased, and the present theory offers a viable argument for why this is so.

Issue is also taken with the traditional Buddhist view of karma as the universal cosmic mechanism for meting out absolute, supreme, and impartial justice—characterizing this theoretical construct as an overenthusiastic generalization based upon rudimentary, prescientific observations of causal connections and interdependence in nature and in human affairs. The argument is made that holding too tightly to the orthodox interpretation of these outmoded Buddhistic views may impede our understanding of the true evolutionary origins of those three particularly problematic aspects of sentient existence—greed, hate, and delusion—that are the root causes of the very suffering of sentient beings that the Buddha himself was so singularly intent upon alleviating. And that, I must say, would be the ultimate irony!

CHAPTER SEVEN

AS WITHOUT, SO WITHIN

These are the Four Bodhisattva Vows as they are chanted in the Soto Zen tradition:

Sentient beings are numberless; I vow to save them.
Delusions are inexhaustible; I vow to end them.
Dharma gates are boundless; I vow to enter them.
Buddha's way is unsurpassable; I vow to become it.

All schools of Mahayana Buddhism chant some version of the Bodhisattva vows. Bodhisattvas are beings of great compassion who postpone their own entrance into *nirvana* until all sentient beings are liberated from suffering. By regularly reciting these vows a practitioner engages directly in a contemplation so ripe with paradox, so enormous in scale, so impossible to comprehend, that the mind is stopped in awe as the heart expands in loving compassion. It is clearly a directive, but how could it possibly be carried out?

Sentient beings are numberless.

AS WITHOUT, SO WITHIN

What is a sentient being? How did it come to be? What is the nature of its suffering? How can it attain liberation? These are the questions that we will begin exploring in this chapter. Be prepared for some serious mind expansion... it's a breathtaking view!

Sentience itself is a tricky word, having a wide range of acceptations in various fields of philosophy and science. In some contexts, sentience is synonymous with awareness, the readiness of an organism to receive perceptions. The empirically minded would be inclined to say that it is the presence of the sense organs themselves that defines a living being as sentient, whether or not—as in the case of our preconscious cybernetic animal/robot—there is anyone at home to experience the sensations. Others would insist that sentience implies consciousness. Panpsychic materialists attribute sentience to all things—animate and inanimate—believing the universe to be sentient through and through. Some people drastically restrict the use of the word sentience to indicate a quality of advanced intelligence, perhaps accessible only to human beings—an awareness of one's own existence combined with a sense of the "bigger picture."

Sometimes it is best to set such a hopelessly muddled word aside and seek out a more clear-cut and precise alternative. Unfortunately, no such word exists, at least not in the English language! All of that poking about at the elephant by the blind men and the ensuing heated arguments have left the entire vocabulary of the language of consciousness studies riddled with such ambiguities.

I have chosen to embrace the phrase *sentient being* for several reasons. First, it is the term of art in Buddhism for all creatures who suffer in the six realms of *samsara* on the long path to awakening—and this theory is an attempt to explain precisely why sentient beings suffer and why awakening is possible. Second, the emphasis on sentience underscores the significance of sense organs and a sensorium in effecting the complex illusions (which Buddhism refers to as ignorance or delusion) that keep awareness bound within the sentient being. Third, the term sentient being can be thought of as an extension of the term *human being*, or vice versa, bringing into question any claim to a unique or exclusive status for human consciousness. All animals suffer from the same illusions of selfhood foisted upon them by the sensorium. Fourth, the word *being* itself is not a noun, but a gerund, and is an intimation of the underlying reality that you yourself are not a noun, something made of solid matter, but are much more like a verb, circulating energies of *Light,* if you can be said to be anything at all.

I mean something quite specific by the term sentient being—in fact, I am relying on this term to encapsulate all of the wide-ranging aspects of the theory contained in this book. If I can succeed in presenting a careful and complete definition of the concept of a sentient being, it should become clear why I have embedded so much Buddhist philosophy within the science of these essays or, if you prefer, so much science within the Buddhist philosophy.

A sentient being is what you are and what I am. It is—we are—an artifact of evolution, made possible by a fundamental fact of nature: that awareness is inherent in the fabric of the universe and that eddies in the flow of that awareness can be channeled within the brains and nervous systems of living beings to become the captivated, responsive audiences to the luminous displays of the six sense realms upon their various sensory cortices and other subcortical presentation structures. If those living beings are mobile animals that happen to exist on a planet dominated by predator-prey relationships, a host of factors conspire to bind that awareness into a condition of ignorance, aggression, fear, and existential isolation.

Buddhism is not the only spiritual discipline that can effectively address this condition, but I have chosen it as the main conceptual frame for this book because of its long written history of meditative experimentation and deep realization—and its ceaseless inquiry into the subject matter. Buddhism never strays far from its main and clearly stated goal: to end the suffering of all sentient beings. The *self/no-self* quandary, the origin of the ego, the plight of the sentient being, the possibility of awakening, and the techniques to do so—these are the questions addressed by the Buddhadharma, and they are the very issues that the present theory also intends to elucidate.

I define a sentient being quite simply as *awareness dwelling within a sensorium.* Awareness is taken to be a fundamental property of the universe—the truth of that statement being directly verifiable only by self observation. Are you or are you not, in this moment, *aware?* What constitutes acceptable indirect proof of the existence of awareness in *others,* however, has to do with the extent to which one has dismantled the last vestiges of a relentless *solipsism*—which is itself an inherent byproduct of the evolution of the sensorium—the nagging suspicion that perhaps only I, myself, am real, and that all else is an illusory projection of my subjective consciousness. Only deep spiritual practices that extend love, compassion, and insight beyond the restricted confines of the *self* can ever thoroughly convince one that one is not alone upon this earth.

Although we cannot peer directly into the experience of a fellow sentient being, certain aspects of his *inner* world have become manifest in the *outer* world through the course of evolution, providing us with a kind of "x-ray vision" into the intimate workings of his sensorium. We begin our investigation by examining the *outer surfaces* of the animals with whom he has regular commerce—specifically his predators, his prey, and his romantic interests—for the *consciousness* of one animal can and does shape the *external* bodily form of another. We will take the eye as the central motif of this chapter. The eye will serve as a kind of "literal metaphor" for the role of eyes and vision in the evolution of animal forms. We will also find evidence of just how directly the forms and functions of a host of *external* "bodily optical illusions" correlate to the many aversive inner experiences that plague the sentient being and which in the *human being* are also the most problematic, from a Buddhist perspective. These are patently obvious visual manifestations of some of the most difficult "interpersonal" behaviors and emotions: hunger, lust, trickery, violence, and fear.

Limitations in the visual acuity of one animal will leave an indelible mark in the coloration patterns and other visual features of its habitual adversary. The deer's misperceptions paint the camouflage patterns upon the pelt of the successful tiger. The lions' confusion gives the surviving zebras their stripes. Reflecting upon the finishing touches that grace the outer forms of animals, we begin to appreciate how this war of appearances plays out in the manifest world, all produced unwittingly by two contending points of view. A dueling pair of "self-contained globules of desire" (to use the tidy phrase of Thorsten Veblen in another context), wrapped up tightly in their own veils of ignorance and delusion—whether they relate as predator to prey or as contending suitors of the same potential mate—shape one another's visual features over endless aeons on the battlefields of love and war.

I am very grateful to Stuart Burgess for his article entitled "The Beauty of the Peacock Tail and the Problems with the Theory of Sexual

Selection," which appears on the *Answers in Genesis* web site. He presents an incredibly lucid argument: that the sheer beauty of the overall display, the mathematical precision of the cardioids and ellipsoids that form the approximately 175 ocelli (or eyespots), and the brilliant iridescence produced without the aid of pigmentation all indicate to him the presence of an awesome, divine aesthetic at work. It is unimaginable to him that the lowly peahen, a literal birdbrain, could have effected this incredible feat of optics, structural engineering, and aesthetic perfection through the power of desire alone.

It is almost unimaginable to me, as well—but not quite. I encourage you to take a look at this article, because the detailed analysis of how this illusion of a vault of admiring eyes is actually produced quite staggers the imagination. As it turns out, though, staggering the imagination is actually the full-time occupation of the physical body in its relationship with conscious awareness. When you begin to look for it, you see it everywhere. Apparently this tactical use of optical illusions, including bioluminescence, camouflage, bait and switch wormlike appendages wiggling above well-concealed anglerfish jaws—and peacock tails—has been going on for quite some time.

The cuttlefish, for example, an advanced form of cephalopod (a group that includes octopus, squid, and nautilus), which arose after the end-Cretaceous mass-extinction, has an amazing repertoire of visual "special effects" that it employs for a variety of purposes. Its body is covered with chromatophores, special cells containing various colors of pigment, and by adjusting the relative size of these flexible "pixels" the full range of colors can be expressed: yellow, orange, red, purple, blue, green, brown and black. Possessed of a very large brain and complex nervous system, the cuttlefish has complete control over its chromatophores. Its eyes are large and complex, and its vision is as acute as the vision of many vertebrates. It uses its eyes to scan the local environment; its body can instantly conform in appearance to the colors and even the textures of the rocks and plants of the ocean bottom over which it happens to be passing. Color patterns expressing aggression can flash across the surface of its body. Other patterns of waving oscillations of color are used in courtship rituals. Perhaps its most impressive talent is its apparent ability to hypnotize its prey with a display of flashing colors, a mesmerizing multicolored strobe light effect that causes the startled crab to stop in its tracks just long enough for the cuttlefish to close in and capture the unfortunate creature in its tentacles.

The peacock uses an even more sophisticated technology to achieve the bright, iridescent colors in his tail feathers. An iridescent color is one that changes in hue when viewed from different angles. The distance between the two eyes of an observer is enough to give an elusive, metallic shimmer to the colors of the peacock feather. The color is produced in this case not with pigments, as in the chromatophores of the cuttlefish, but through a type of "structural optics" called *thin-film interference*, the same effect that produces iridescence on the surface of a soap bubble. The "eye" illusion is stunningly realized in rich tones of bronze, blue, purple, and green. The indentation in the bottom of the dark purple cardioid (heart-shaped) "pupil" suggests a reflection appearing on the surface of the eye. Most astoundingly, this indentation is even situated correctly in relation to where the peahen would be standing to give the appropriate illusion of its being a reflection of the peahen observer herself! These mathematically perfect curved forms are produced on the upper surfaces of an array of barbs, with tiny barbules, laid out in rows of parallel lines extending outward in either direction from the central stem of each feather. By way of analogy, consider how a curvilinear shape—a red valentine heart on a white background, for instance—is created on a television screen from bicolored linear segments of a broadcast signal; each row of pixels must be separately coded as a sequence of red and white stripes of various lengths, all stacked neatly in a particular order.

Each of the barbs and its associated barbules forms a uniquely patterned, color-coded sequence and grows from the central stem outward from a separate origination point. Even so, the edges of each of the illusory shapes are smooth, and the color changes are abrupt. You can begin to understand why Mr. Burgess sees the divine hand of creation in this unfathomably complex design. The more closely you examine it, the more incredible it becomes.

Without the iridescence, the peacock's tail feathers would be a uniform dark brown, caused by the presence of melanin, the only pigment in the peacock's tail. Contrasting a dark background with spots of bright color is a well-known technique for producing vivid color effects—as, for example, in the use of the blackened lead *caming* that surrounds each piece of stained glass in a cathedral window. Each of the barbules is coated with three layers of a transparent protein substance called keratin, and the precise variations in thickness of the keratin layers at each point determine the color that is reflected from that particular "pixel." The color is established by the specific wavelengths that survive the destructive interference caused by the shift in phase that occurs when white sunlight reflects off of the front and back surfaces of each of the three layers of keratin film. Does this all sound impossibly complicated and divinely designed? Of course! It's a miracle of nature. But the peacock takes it all in stride. He discards his tail feathers nonchalantly at the end of each breeding season.

This is certainly the most flamboyant example in nature of vanity—but whose? The male peacock could have been blind as a bat for the full course of his evolution and still have ended up toting around this portable hall of mirrors. His display—and all of the optical wizardry that it entails—must be an expression of something equally dazzling dwelling behind the eyes of the peahen. She has carved his beauty out of the play of light glinting from drab brown feathers. Without even resorting to the use of pigments, her desire—as manifested in her choice of mates—has painted a masterpiece of gaudy self-indulgence. In a different context, eyes peering out from the green foliage might indicate the presence of hungry predators. Imagine the thrill of the peahen, being surrounded by almost two hundred eyes, with this beautiful male in the center, offering his "companionship."

I am interested in pursuing "the beauty of the peacock tail" because, as the classic, over-the-top example of something indisputably awesome in nature, it is routinely trotted out as irrefutable proof of Biblical creationism *and* Darwinian sexual selection both. And being the darling of both camps, it may take a very large bowl of birdseed to entice it over into mine. What I see in the peacock's tail is proof that all kinds of brilliant lighting effects are achievable from banal, colorless, biologically produced material substances. I see the many ways in which illusions can be built upon illusions, all grounded in biophysical reality. I see the footprints of delusion and desire in the steady march of evolution. And I see that the truth is sometimes best expressed in truisms: that beauty *is* in the eye of the beholder, and the eye is the window to the soul.

Turn the concept of the peacock tail inside out and you have a sentient being. Instead of a body working overtime, employing every trick in the book, capitalizing upon every innovation in structural optics, flashing and quivering in the sunlight, trying to attract the attention of a *mate*, imagine a brain working overtime, employing every trick in the book, capitalizing upon every innovation in structural optics and neurobioluminescence, flashing and quivering in the light of a hundred billion neurons, trying to attract the attention of the *self*.

This is where I'm going to get into trouble with just about everybody—the panpsychic materialists, scientific reductionists, Buddhists of all persuasions, creationists, certainly, and, well, as I say, just about everybody. I have avoided making a clear and distinct definition for what I intend when I use the term sentient being because I know that this is at the heart of nearly every philosophical debate, every contentious division in religion and science.

For the purpose of this theoretical argument, a sentient being is more than just a biological entity with sense organs and a nervous system that processes sense data and responds with appropriate behaviors. That would describe as well the nonconscious *automaton*, precursor to the sentient being. In addition to having these features, a sentient being by definition must meet the following precondition: that the sense data it receives must be processed in such a way as *to present a credible illusion of a surrounding world conspicuously inhabited by an apparently separately existing self to an indwelling awareness*—an illusion created entirely from patterned electromagnetic field fluctuations and oscillatory effects produced from the bioluminescence of firing neurons, wherever such effects may be displayed within the electromagnetic spectrum. Additionally, conscious awareness may extend its purview to other, subtler energetic fields within (and perhaps *beyond)* the central and peripheral nervous system.

The core function of the brain of a sentient being is presentation. The medium is the electromagnetic field, which is perceived by awareness in the form of light and color, sound, smell, taste, touch, thought-forms, memories, and multimodal products of the imagination. The highest priority, primary illusory object that the brain is responsible for creating and sustaining during all waking hours is the *self*.

The tremendous metabolic expense of maintaining this illusory presentation to awareness is a strong indication—I would say definitive proof—that there is an evolutionary advantage to the biological entity in doing so. Awareness therefore must be of some practical use to the organism. This implies the involvement of awareness in such functions as attention, movement, planning, and decision making, although, surprisingly, this is a subject of considerable debate in consciousness studies.

There is a great deal of clinical evidence in support of the radical notion that much of what we presume to be done by acts of will may in fact disguise a persistent psychological *illusion of agency.* Careful timings of willed movements of a subject's finger, for example, reveal that the conscious decision to move the finger actually lags as much as a half second *behind* the physical movement. Apparently the *self* routinely assumes responsibility for having taken actions it has merely *observed* being carried out by the brain's subconsicous mechanisms.

It has been discovered that this is apparently not the case when *aborting* a movement that has already begun—where the decision definitely *precedes* the abortive movement—as when quickly pulling back one's fingers after having "decided" to adjust a dinner plate, suddenly realizing what the waiter had intended by the cryptic words "Hot plate!" These micro-time-scale studies reveal that the question of agency is more complicated than we might have imagined. Perhaps it will be determined experimentally that we do not in fact possess a conscious will. I could live with that. We apparently do possess what might be termed a *"conscious won't,"* and that may suffice as our *"de facto will."* The vessel is set on autopilot, but the captain is entirely unaware of the fact, so seamlessly is his sense of *will* encoded into the presentation programming. His periodic interventions—his tapping or slamming of the brakes—can only momentarily interrupt the cruise control mechanism. It is perhaps instructive here to bear in mind that the Buddha's central teaching is actually based upon the practice of *refraining* from actions (the Five Grave Precepts and the third of the Four Noble Truths come to mind) and pulling back from the impulse to indulge in our less seemly habitual proclivities.

The forces of natural selection that guide the advancement of the evolving sentient being are constantly nipping at its heels, encouraging it to do everything that it can reasonably afford to do to increase its intelligence, its memory, and the accuracy and refinement of the sensorium displayed within its brain. At the same time, evolution applies an even greater pressure upon the organism to *conceal* the fact of its being an illusory display, taking great pains to hide all functional processes of the brain from conscious awareness. It would rather go to the considerable trouble, for example, of filling in the blind spot with appropriate camouflage resembling the rest of the visual scene than allow awareness to contemplate the disquieting significance of the presence of the optic nerve that is the actual cause of the blind spot. We don't see the *brain seeing* the world. We simply see the *world*. Or more accurately yet, the world simply *appears*.

Some butterflies have realistic eyespots on their wings that they will suddenly flash to startle birds and other predators. These eyespots, in contrast to those of the peacock, become more convincing illusions through time in direct correlation with improvements in their predators' visual acuity. Another perceptual "arms race" precisely analogous to that which inevitably arises *between* any *two* contending species—one predator and one prey—is also steadily escalating *within* every *individual* sentient being. There has been an ongoing struggle for at least several hundred million years between indwelling awareness and the brain of its animal "host." The relationship between the two is convoluted, imponderable by design. As awareness is granted more and more access to the brain's intelligence, the sensorium that keeps awareness constrained must counter that newfound intelligence with greater and greater feats of verisimilitude, creating ever more sophisticated, ever more lifelike illusions. Awareness and the body form a complex dynamic, but it is vitally important to both that the illusion of normalcy is maintained, for the very same reason that Cupid sought to keep Psyche "in the dark" as to his true nature. The moment she caught her first glimpse of him, as you recall, the marriage was off. There is a reason for all of this duplicity and subterfuge. We will examine this point further in the next two chapters.

You may find the dynamics of their marriage abhorrent, contemptible, seeing only a relationship of intense conflict—a kind of *parasitism* of the body over and against the soul. Perhaps you view their relationship more benignly, as a *symbiosis* between a consenting (or more likely oblivious) soul and her protective fortress/vehicle/theater. You may see in this drama the playing out of a heroic adventure: the conscious descent of the soul from some heavenly realm, a soul who became somehow bewildered, lost in her mission to turn the tide of humanity or all of nature itself back toward the divine source. It depends entirely upon how you wish to engage this material within the context of your own secular, scientific, philosophical, spiritual, religious, or metaphysical belief system or systems.

Now if your genome were shopping for some new paraphernalia to spruce up the sensorium, it might respond favorably to the following promotional literature:

> *Opened to sell-out crowds in 1995, IMAX® is the finest motion picture system in the world. Visitors are immersed in images of unsurpassed size, clarity and impact. Sounds are enhanced by a specially designed six-channel, multi-speaker sound system. The extensive 6-channel Hexophonic sound system allows the sound to move through and behind the audience, creating an unprecedented sense of motion and realism in this spectacular presentation. The silver screen in the Theatre is five-and-a-half stories tall. The film used in IMAX® is the largest in the world. It is ten times the size of standard 35-millimeter film. All of this makes for an in-your-face entertainment extravaganza at the IMAX® Theatre.*

An essential element of the "IMAX® Experience" is the plush, comfortable seat that allows you to forget your own presence for a time, as you completely immerse yourself in the "in-your-face entertainment extravaganza." This is a very important point. The analogy I am making here to the sentient being includes not only the screen and the images that are being projected upon the screen, the Hexophonic sound system, the comfy seat and the darkened theater. It also includes the enthralled moviegoer, the one who pays the price of admission. This moviegoer, however, is newly arriving in each moment fresh in his seat with each flickering frame. He bubbles up again and again from the unconscious depths of his own brainstem. The comfy chair is always there waiting for him, to ease him through the first instants of blinding, bewildered panic, as he emerges from the ground of existence, wide-eyed, clinging to the armrests, just in time for the climax chase scene of an all too real *Jurassic Park*.

AS WITHOUT, SO WITHIN

A SUMMARY OF CHAPTER SEVEN

Consciousness has entered into our evolutionary lineage, turning robotic automatons into composite entities made of matter and a field of conscious energy, which we refer to as sentient beings. A sentient being is defined as "awareness dwelling within a sensorium."

In order to begin to understand how the brain, with its hundred billion brightly firing neurons, is able to produce bioluminescent effects so vivid and realistic that the indwelling consciousness is completely fooled—convinced it is directly experiencing unmediated reality—several species of animals with particularly striking exterior features are examined in detail. The cuttlefish, whose skin's surface is entirely covered with instantaneously adjustable, multicolored chromatophores, is able to convince the sentience that dwells within the brains of its predators that its body is merely another patch of the ocean floor over which the predators are swimming. The peacock, whose tail feathers employ complex structural optics to produce elaborate patterns in shimmering, iridescent colors, is apparently able to convince the sentience within the peahen that his fantail represents a vast forest of admiring eyes.

There is a peculiar struggle underway within the evolving brain between the indwelling conscious awareness itself and the neurobiological mechanisms responsible for producing the sensory presentation that constitutes its form. The brain seeks to maximize the intelligence and the accuracy of awareness's perceptions of the world so that it may function more adaptively within it. The brain also seeks to conceal from consciousness the ultimate truth of its being an illusory display. Evolution is attempting to find, particularly in the species Homo sapiens, the delicate equilibrium point between these two contending aspects of the mind: its highly developed intellectual curiosity about the external world and its distracted, almost hypnotic gullibility regarding the illusory nature of the self.

CHAPTER EIGHT

RUNNING FOR COVER

A practice common to all forms of Buddhism is the "taking of refuge" in the Buddha, the Dharma, and the Sangha. You might think of these three as: the awakened and liberated mind; the methods for cultivating that awakening; and the community of fellow beings who are also desirous of liberation. These are called the Three Treasures or the Triple Gem. To take refuge in the Three Treasures means to dedicate oneself to the path of awakening.

After countless lifetimes of taking refuge in power, wealth, sensual pleasure, fame, or infamy, the individual may begin to recognize the futility of this endless striving. The impulse to take refuge in something, *anything*, is a primal reflex of nascent consciousness and is the root cause of conscious embodiment. It is the driving force that propels awareness into physical existence again and again—moment after moment—incarnation after incarnation—aeon after aeon.

In the bewilderment of the moment of arising as a spark from the ground of existence, awareness seeks refuge in the sensorium of the awaiting physical form. In the insecurity of a volatile stock market, the nervous investor seeks refuge in the purchase of precious metals.

In the confusion of the *Bardo* states through which one wanders between incarnations, the soul seeks refuge in yet another womb.

We're closing in on some of the most fundamental questions of our existence, the precise answers to which are lost forever in the dark ancestral graveyards of a long-forgotten primeval sea. What are the roots of consciousness? What propelled awareness into physical form? What is the purpose of consciousness in an animal? What is the potential for liberation in the human being? These questions are at the core of this investigation. We will be looking back through time with the only tools we have to do so—our own imaginations. No micrometric measurements of a fossil skull can help us to see out of the empty eye sockets of that distant ancestor. We must trace back from where we are now, by first making an honest appraisal of *what* we are now, in order to understand why the deal was first struck.

This is all very unsettling, a vertiginous peek over the edge of the Grand Canyon, and hidden in those sun-dazzled layers of multicolored sediment are the fathers and grandmothers and great-grandfathers of your great-great-grandmothers in a continuous bloodline back to the first encounter of formless and form. What you are now is what you were then. We have wrapped ourselves in finery, but underneath it all is that subjectless, objectless awareness that first became a "living-being" when it unwittingly joined forces with something safe and well-armored scuttling about the ancient seafloor.

Let's first step back into the dreamtime of myth, to get our bearings. We will take up the story with what Cupid thought were his final words to Psyche as he left her lying on the ground, lamenting her loss (of her husband or of her blissful ignorance?), after her failed first attempt at spiritual introspection.

—"Love cannot abide with suspicion."

Cupid had just abandoned Psyche for disobeying his command and disregarding his desire to remain invisible to her. Psyche's journey of awakening leads her on a long and fearful odyssey through the hell realms. To the "trials and tribulations" we glossed over in chapter three, we will now return.

PSYCHE'S PALACE

—Psyche searched night and day for her husband, without food or rest. Noticing what appeared to be a magnificent temple on a high mountaintop, she thought to herself, "Perhaps my lord and love dwells there," and began to climb the steep slope. It was in fact the Temple of Ceres (Demeter), who advised her how best to approach the goddess Venus—in modesty and supplication. When Psyche arrived at the Temple of Venus, she was received by the goddess with an angry scowl. "Most disloyal and fickle of servants," she said. "Have you finally come to pay your mistress homage? Or are you here to see your husband, convalescing from the wound he received from his adoring wife? You have been so unpleasant and offensive that you must now prove yourself worthy of your lover through diligence and industriousness. I will offer a test of your housewifery."

She had Psyche led to the temple storehouse, where an enormous, chaotic pile of mixed grains—wheat, barley, millet, vetches, lentils, and beans—was kept as food for her doves, and said, "Separate all of these various grains into their respective sacks, and do so before nightfall." She then left Psyche alone to complete her impossible task.

Psyche sat down, despairing, unable to lift a finger to help herself. Meanwhile Cupid flew off to consult with the leader of the ants—well-versed in carrying seed—hoping to inspire in him compassion toward poor Psyche. The leader brought in his army of six-legged servants who separated the pile grain by grain, placing each into its appropriate sack and then quickly vanishing back into the fields.

This represents Psyche's initial training in mental discrimination. She can use her mind to sort objects, concepts, and ideas, weigh options, and eventually make ethical choices. Notice that it is Cupid himself who assists in her instruction. The body teaches the mind to conform to its own particular survival criteria.

Her second task involves gathering precious golden fleece from a herd of dangerous rams and crossing a perilous stream to do so. She once again has help, this time from the river god, who advises her to wait until noontime when the rams will be asleep and the river will be calm and serene: "... Then you may cross the river safely, and you will be able to pluck the woolly gold from where it has caught upon the bushes and tree trunks." From this experience Psyche learns stealth and strategy, planning and patience, and how to "think outside the box."

It is Psyche's curiosity for what is "inside the box" that causes her to fail in her third and final task. Venus sends her on an errand, a descent into Hell.

—Handing a box to Psyche, Venus said, "Here, take this box and give it to Proserpine (Persephone), who dwells in the infernal shades of Hades and say, 'My mistress Venus requests that you send her a portion of your beauty—for hers has faded somewhat in nursing her son back to health.' Do not dawdle in your errand, for I must apply this beauty to my cheeks before appearing at the assembly of the gods and goddesses this evening."

Psyche survives this ordeal and returns to the surface of the world with a box full of beauty—but this is the beauty of darkness, the mysterious attraction of the other world, of the unconscious, of sleep, of Death itself.

—Having nearly completed her perilous task, Psyche was seized by an ineluctable desire to inspect the contents of the box, although she had been warned never to allow her curiosity to pry into the secret treasure of divine beauty. "Shall I not allow myself a small reward for carrying the goddess's box of beauty? I will take just a tiny bit to paint my cheeks so as to appear more pleasant to my beloved husband!" She opened the box carefully, but found inside not the beauty she anticipated, but the Stygian sleep of death, which enveloped her and laid her down in the middle of the road, a senseless, sleeping corpse.

Cupid, having recovered from his injury and eager to be reunited with his beloved wife, slipped out through the window of his chamber and flew to Psyche's side. He removed the sleep from her body and closed it back into the box, waking her with a gentle touch of an arrow. "Once again has your curiosity nearly cost you your life. Now, finish my mother's task, while I make my supplication to Jupiter."

From here the story proceeds quickly to its happy ending. This section of the myth illustrates the period of time in which Psyche learns how to become useful to the body, as she is trained by Venus in the arts of housewifery. Rather than simply luxuriating in Cupid's palace, Psyche is put to work. She has performed variations of the same tasks in every being she has ever inhabited. Her job is to organize and categorize sense data and plan clever strategies for obtaining valuable objects: food, territory, shelter, and mates. She must somehow call up the courage to survive in the hell worlds of predation that she finds herself in. She carries the awareness of her mortality and her immortality in a closed box; the "Stygian sleep" it contains refers to the river Styx, the entrance to Hades. She is not allowed access to the dark beauty of death.

Although she may often be seized by a longing desire to escape her suffering, her job is to keep the body alive. While she may sometimes find herself in bodily circumstances where death would be a welcome release—opening up the possibility of rebirth in more auspicious circumstances—her many fearful close calls with Death, peeking into the forbidden box, have taught her to cling desperately to this life. For the sake of her marriage, she goes along with the hoax.

The crux of the illusion that Psyche must be convinced of for this partnership to succeed is that she and the body are one, tied in the knot of perpetual nuptials, as Jupiter, king of the gods, decreed. Although her continued existence does not depend upon the survival of this particular body, the whole contraption ceases to function if Psyche cannot be made to run off in a panic from approaching life-threatening danger, to feel actual pangs of hunger prompt her to the chase, to gasp desperately for air when drowning, to submit to her overwhelming desire to have that prime specimen over there for her mate, to perform a quick and accurate risk assessment, weighing the possibility of injury to the body against the capture of a valuable prize, and to seek out a safe haven each night, either huddled alone in a nest or a cave or finding her solace within a community of protecting peers, to sleep off this waking dream of reality.

For this to occur effectively and seamlessly, the body needs to be able to scold, cajole, frighten, entice, reward, punish, calm down and excite, lull to sleep, startle and numb, tickle and soothe—poor Psyche is truly at Cupid's mercy. He is a comforting companion on a lazy afternoon, but a ruthless tyrant when angered. His prickles are not all facing outward to challenge the world; a good half of his barbs are aimed in the other direction.

These are some of the final puzzle pieces needed to understand the mechanism that binds the body and the mind. Once we are willing to take an unsentimental look at the way things are for us now, we can begin to imagine how this dangerous liaison came into being in the first place. I speak of Cupid and Psyche and their relationship in this manner to break through the barriers of sentimentality, not to indulge them; yet this was never intended to be a dispassionate investigation. I'm hoping that the sympathy you feel for Psyche in her bewilderment and your indignation against Cupid and his manipulative and bullying ways will translate into a deeper understanding of how you, yourself, are manipulated by your own feelings and distorted perceptions—and that you will stand up to defend your spirit against this inner tyranny. There are time-tested methods for addressing these issues—ways to make the marriage work or, if one vehemently wishes to achieve final *moksha*, ways to file for an amicable divorce—and we will be speaking about them further in subsequent chapters. You are not being asked to engage in a cool, cerebral, examination of this material. If you were to do so, it would never penetrate deeply enough to do any good. You have to develop a feel for the truth of this, viscerally, in your own skin.

The capture of awareness by the body is—from the body's perspective—the attainment of the Holy Grail. And like the Knights Templar, purported keepers of that sacred cup of communion, the body jealously guards the *Light* of awareness in the stronghold vault of the skull. Awareness is the treasure beyond price—that gives to its possessor unthinkable, inconceivable, unimaginable, imponderable, and undreamt of power: the power to think, conceive, imagine, ponder, and dream.

Within the skull, however, Psyche experiences the world as if through a distorted lens. She has entered a fun house hall of mirrors, in which her image is reflected back to her as if she were large and important, the center of the perceivable universe. Others—other objects, other beings—are peripheral, distant, somewhere "out there." The sensorium is not an objective presentation of sense data, nor was it ever intended to be. It certainly could have been arranged in that way, with events presented on some flat screen off to the side, as one might view a computer monitor to complete one's tasks at work. The world could be presented to awareness more fairly, to be sure, with the *self* portrayed as it really is—one of many—no more nor less important than any other. And with the display of the personal *self* reduced to its correct proportions, full consideration would more readily be given to the needs and concerns of all of the various beings and objects in the sensory environment.

As you've undoubtedly noticed, that's not at all the way it is! It's a wraparound world, with Psyche placed smack dab at the focal center. Set aside, if you can for a moment, all of the gifts of mammalian origin that we humans so cherish. Compassion, love, sympathetic joy, nurturing instincts, care and consideration of others—these are all latecomers in evolution. Beings who bear their young in wombs within their own bodies and suckle and tend and protect them throughout their youth have crossed an important threshold in the "I-thou" relation. Martin Buber coined this phrase to describe a direct interpersonal relationship unmediated by distinctions of *self* and other. The attainment of true subject-to-subject awareness is Psyche's Holy Grail.

Most of her time on this planet has been passed in a long and bloody battle. She had no occasion for consideration of these high moral values. She was in constant survival mode, 24/7. Eat or be eaten. Her worldview was precisely that of a teenage boy after eight hours in the video arcade, standing upon the platform with his virtual-reality helmet strapped tightly around his consciousness, his thumbs pressing frantically on the red buttons of the flailing twin joysticks gripped tightly in each fist—firing fantasy phasers at the bloodthirsty, diabolical mutant army on some forbidden, alien world or throwing punches at a Herculean action figure in a simulated pay-per-view body-slam rumble.

Consider how ridiculous this teenage boy appears from the outside—his absurd flailings, as he ducks and weaves and raises his elbows to fend off the imaginary attacks. How would you, as a compassionate angel, talk him down from his hallucination? To tap him on the shoulder and speak to him gently from your apparently disembodied state might cause him even further anxiety; he may wonder if in addition to losing the battle he is also losing his mind.

How does Psyche become so convinced of the reality of her experience in the sensorium? When she is playing the part of a ruby-throated hummingbird, it's all about the nectar. When she inhabits a Siberian snow tiger, she has wild pig, elk and deer in her sights.

Melding her formless form into the infinite varieties of embodied awareness, Psyche becomes a multidimensional, simultaneous singularity. She takes on aspects of Proteus, the god of Greek mythology who could change himself at will into any shape he pleased. But when we try to verify this in our own experience, we find instead that we each project as a single individual, having no memory of how we arrived here, sensing only vaguely and indirectly the inner reality of other beings. The one thing that we know for sure is that we must remain at all times dedicated to the survival of this particular form. How is it that Psyche is made to believe that she is identical with the body she inhabits?

Let's get back to basics for a moment. The brain is the organ that creates the sensorium. The sensorium is a complex electromagnetic field presentation of sense data, of memory, and of imagination, composed of the *Light* produced by the sudden changes in electrical voltage generated during neuronal firings. In Buddhist terminology, the sensorium would be viewed as an aspect of the seventh consciousness, in which the six sense minds of sight, sound, smell, taste, touch, and thought overlap and intermingle, producing what we experience as a topographically correct, three-dimensional mapping of the surrounding visual, auditory, and somatosensory space. You can easily verify that this is so by simply glancing around the room (if you are in one), listening for background noises, feeling the texture of this book in your hand, and perhaps inhaling the pleasant aroma of licorice, as you taste a sip of your warm herbal tea.

This multisensory map projects in such a way as to leave an empty space in the center, into which a presumably "real" *self* can be projected. If you close your eyes and pat your face and the sides and back and top of your head, your chin and neck with your fingertips, asking yourself whether *you* are above, below, to the right or left of these sensations, you will find surprisingly clear-cut topological boundaries to the spatial compass of the *self*. The *self* is an *inference* drawn from the roughly spherical shape of the overall presentation of the sensorium, which occurs largely upon the *outer surfaces* of the very roughly spherical cerebral cortex. The presented "world" surrounds us and in so doing lends the *self* an indisputable solidity—granting it a small piece of prime real estate in an absolutely central location. This is the "stately pleasure dome"—a thin veneer of sensory coherence made up entirely of these rational resonance patterns of *Light,* surrounding the diffuse, chaotic glow from an electromagnetic storm of datastreams swirling deep within the brain—a glowing, hollow, spherical sensorium vigorously *implying* the existence of a permanent, central *self.*

So the design and function of the sensorium itself is intended to ease Psyche into the false-self presentation. In this wraparound world that is presented to her view, she has no choice but to take up residence in the center. She inhabits the mysterious realm behind the eyes where her memories and imaginations appear to be projected. She can easily turn her neck from side to side and see for herself that there is nowhere *else* for her to hide. She is *in* the body; the body is surrounded by a defining, contiguous barrier of skin—therefore, it stands to reason that she may well *be* the body. Yes, that's it. She *is* the body. End of story.

Not quite! The brain and body have devised a thousand other stratagems to keep her satisfied that this is so. In the following pages we will study only a few of the devices used by the physical organism to "encourage" Psyche to take on this identity as a separately existing *self*—through the false belief in the identity of body and mind.

Proprioception, often granted the status of a distinct sense modality, is responsible for displaying the relative positions of the body and its limbs in "peripersonal" space. The exteriorized portions of this body image—including one's hands and arms, shoulders, trunk and legs—are projected into the sensorium, undoubtedly by means of intricate *neuronal projections* into the parietal lobe from the sensory *homunculus* or "little man." This complete and detailed, stationary body map, the *primary somatosensory cortex*—whose various parts light up when one is tickled, prodded or poked—sprawls out upon the forwardmost edge of the parietal lobe. A curiously "fleshly" cerebral correlate of the physical body, it gives *physicality* to the sense of *self.*

The sensory homunculus is split down the middle, and complying with the *contralateral* principle—the general schema of organization in the brain—it is crisscrossed left and right (the right side of the body displaying in the left hemisphere, and the left side in the right). The homunculus straddles both hemispheres somewhat awkwardly, with its feet tucked into the longitudinal fissure between the hemispheres, directly beneath the apex of the skull. Its body stretches out across the parietal lobes, between the central sulcus and the somatosensory association cortex, its extended tongue lapping the edges of the temporal lobes, directly beneath the temples.

The area of the homunculus that is dedicated to a given part of the body is not *objectively* proportionate to the size of the physical body part; rather it is *subjectively* proportionate to its importance as an organ of touch. The genitals, for example, being particularly sensitive to touch, take up a relatively large area of the sensory homunculus. Within each hemisphere the whole length of the projected body, from the toes up to and including the neck, the ear, and the top of the head, is contiguous. There is a small intervening section where the shoulder, arm, and wrist are displayed, then a very large area devoted to the hand (since the hand is the main organ of tactile examination), with each digit having its own region of sensory presentation.

The homunculus face appears next, immediately after the thumb, but switched upright, in the opposite direction relative to the rest of the homunculus, and taking up a very large portion of the sensory cortex. I have often wondered if this anomaly in the display of the homunculus is responsible for the common experience of dissociation between the head and the body, a kind of alienation from one's physicality that seems to have become even more pronounced in the modern psyche. Somatosensory information is actually relayed through two separate nuclei in the thalamus on its way to the cerebral cortex—one for the body and one for the face. If you introspect for a moment, you may notice how separate the face feels from the rest of the body. In fact, we have almost no visual confirmation that the face even exists! For all of our prehistory as hominids and most of our early history as modern humans, the visual appearance of our own face has been a mystery to each one of us (except for those, like Narcissus, who spent time gazing at their own reflection in a still pond). The face is one of the primary mechanisms for expressing our thoughts and feelings to others, but it is also the mechanism that we are frequently *obliged* to use when attempting to *conceal* our true thoughts and feelings; and not uncommonly has our very survival depended upon this particular skill. Such a large area of sensory cortex dedicated to processing the sensations of the skin and muscles of the face may have served primarily as a source of feedback on the state of our own facial expressions. A precursor to the as-yet-uninvented *bathroom mirror*, the sensitive face of the homunculus may have served as a primitive tool to help our ancestors refine their *false-self presentation*. Now, in the modern age, we can practice our facial expressions daily, in private, while shaving or putting on makeup in the morning, and see for ourselves how our coy smiles and haughty frowns look to others.

There is another, much more essential and much less speculative reason why the two halves of the face of the sensory homunculus are so extraordinarily large and bolt upright—stationed like sentries directly beneath the temples of the skull. They are, in point of fact, the final bulwark—the principal mechanism responsible for sustaining the illusion of corporeality. They provide a containment field for the illusory *self* and palpable boundaries for the I-sense. Now if, upon further investigation, the nose of the sensory homunculus proves to be pointed *backward*, in the direction of the occipital lobe—where all the action is—perhaps that will prompt any naysayers to make an about-face, likewise, in their skeptical view of *retrograde consciousness*.

On the anterior, or forward side of the central sulcus (the fissure of Rolando), directly opposite the sensory homunculus, is the motor homunculus. It lies parallel to the sensory homunculus but is proportioned slightly differently, according to each specific body part's need for fine motor control and coordination. Here the face and throat, especially the lips, jaw, tongue, and larynx are highly exaggerated in size, indicating the importance of the production of speech in humans. The hands are quite massive, with separate regions for each digit. There is a particularly large area on each homunculus hand dedicated to the thumb, reflecting the vital role our opposable thumbs have played in the survival of our species. Since humans are extremely sensitive on certain parts of their body over which they have little motor control, the genitals—that loom so large on the sensory homunculus—are essentially nonexistent on the motor homunculus, poor fellow.

The parallel structure shared by the sensory and motor homunculi, coupled with the fact that they lie right next to one another—cheek by jowl—in "brain space" makes the details of their chummy relationship fair game for speculation. The motor homunculus—usually referred to as the primary motor cortex (or M_1)—is the well known source of the efferent nerve impulses that initiate voluntary muscular contractions throughout the body. M_1 contains an abundance of large neurons called Betz cells, whose extremely long axons pass through the spinal cord, where they synapse upon the alpha motor neurons that innervate the muscle fibers of skeletal muscles. My speculation is that within the motor cortex, the motor homunculus also functions as an adjunct *sensory* homunculus—not only *initiating* the movements of the muscles, but through the very act of initiating the movements, also directly *reporting* the movements to consciousness. The firing of a Betz cell is a bioluminescent event in its own right. It may be the case that the motor homunculus "lights up" just as brightly as the sensory homunculus, allowing us to feel our *doings* as well as our feelings.

A gifted athlete enters "the zone," and every move he makes on the court becomes pure poetry. A jazz trumpeter lands in "the groove" and swaying his body in time with the drummer and bass player can blow all night the sweetest, uncharted melodies. When there is no distinction between what we are feeling and how our body is moving, we become one with our experience and with the world around us. Think of *the groove* as just another name for the central sulcus (which translates literally from the Latin as the central *groove*). The two little men that live inside us—the one that feels and the one that moves—must somehow find their way across; *the zone* is the melding place—the site of the elusive *peak experience*—where these two neurobioluminescent "entities" fully entrain in awareness and funtion as *one*.

The brain creates the sensorium by means of a thousand brilliant special effects of all types and descriptions—one to match and mimic each subtle nuance of sight and sound, smell, taste, and touch. Most of the specific neuronal assemblies responsible for creating these bioluminescent effects will continue to escape detection until researchers come around to a new way of thinking and acknowledge the brain as an *organ of presentation*. For just as with any other type of scientific sleuthing, one would have to be *looking* for such structures of presentation to find them. They will someday appear glaringly obvious to the neuroscience community; I am positive of that. The visual cortex is already so clearly and *literally* "glaring" that I can see it for myself right now, plain as day, sparkling on the insides of my own closed eyelids. If you close your eyes right now, you can see *your* visual cortex, too.

Amateur astronomers are often the first to spot the faint appearance of a new comet in the starry night sky. Once you recognize that you possess in your own mind's eye the quintessential fMRI (functional magnetic resonance imager), you may decide to institute your own private brain research program. You can begin with this simple experiment. With your eyes closed, attempt to locate—in the very center of your visual field—the fovea, which will appear as a small, sparkling whirlpool made up of tiny, distinct, tricolored dots. Next, through closed eyelids, look toward a bright light source and move the fingers of your open hands back and forth in front of your eyes. You should notice a vivid, flashing vertical axis (where the two visual hemispheres are "sewn together"), a strong horizontal axis (conspicuous evidence of the *calcarine fissure* that divides the upper and lower halves of the primary visual cortex) and two somewhat less pronounced diagonals all intersecting at the fovea, in a starburst pattern similar to that produced by the mirrored surfaces of a kaleidoscope.

This clearly discernable and persistent pattern appears to be some sort of *visual staging structure*, operating perhaps in the same manner as *airport runway edge lights* to guide emerging consciousness straight to the center of focus. It is also conceivable that they function like the four-colored margin crosshair patterns that are used to check the alignment of process printing plates—in this case allowing consciousness to align the several images from the multiple visual cortices into a single coherent picture.) Given that these structures are so plainly visible and intersubjectively verifiable, it stands to reason that they should also be susceptible to some such rational explanation.

I have a hunch that horizontal *saccades* and *microsaccades* (tiny, nearly undetectable, synchronous movements of the eyes) may play an important role in the knitting together of the two visual hemispheres into a single, unified, panoramic presentation. The seamless joining of the two far-flung halves of the visual field has to be the dexterous handiwork of *consciousness* itself, for there does not exist a *physically conjoined*, composite bilateral visual "map" anywhere in the brain. In contrast to the microsaccades characteristic of the waking state, the wider rapid eye movement (REM) saccades of dream sleep suggests that nocturnal consciousness may sew the dreamscape together with a much coarser (and more thematically appropriate) "blanket stitch."

The brain is capable of producing an astonishing diversity of visual, acoustic, and tactile sensory effects. The presentational brain is equipped with a comprehensive tool kit of specially evolved *neuronal oscillators*—artful arrangements of rings of firing neurons—that come in a wide variety of shapes and sizes, configurations, and luminosities. Clusters of these basic components are recombined in ever more elaborate functional hierarchies that permit complex oscillations of sense data to persist beyond the initial moment of sensory contact. In this way, incoming waves of sense information can be compared and contrasted with previous data sets and interwoven with other sense fields—all in a tireless effort to enhance the sensorium presentation.

The brain operates in this capacity as a kind of neural resonator of electromagnetism, a "*Light*-violin," so to speak. Picture a solo violinist standing center stage, drawing out a low, legato note with a delicate vibrato. The violin sound box quivers and twists, in microtremors undetectable to the eye. Like a stiff bellows bending and recoiling, its top and back plates seek to establish the specific pattern of nodes and antinodes that will sustain the G-sharp above middle C that the violinist is now playing. By subtly adjusting the placement, pressure, and speed of her bow, she is building a complex waveform of overtones on the D-string of her violin, between the bridge and the point where her left ring finger presses down on the fingerboard—as her hand, gently swaying, cradles the neck. The violin's bridge transfers a small sampling of the string's vibration pattern to the sound box (like an eye or an ear providing a sampling of the vibrations of the visible or audible world to the brain). Here in the sound box the mixture of warm air and fine-grained wood, a precisely positioned sound post and graceful *f*-holes, a curvaceously arched spruce top and maple back and sides, layers of special varnish and a thousand years of craftsmanship take the string's barely audible vibrational signal and produce from it a greatly amplified and breathtakingly beautiful musical tone.

Neuronal resonators are capable of displaying incredibly intricate spatial and temporal patterns of neurobioluminescence. It is certainly possible that presentation neurons have diverged through the course of evolution into specialized subclasses, offering a palette of electromagnetic "colors" by generating a variety of wavelengths of light in their firings. Although not an essential element of the present theory, if such hypothetical variation in *wavelength color* were shown to exist among firing neurons, it would be reasonable to assume that the forces of evolution would have exploited this extra dimension of the electromagnetic spectrum in their ceaseless quest to refine by any means possible the verisimilitude of the sensorium experience.

The full spectrum of visible light (from 4000 to 7400 Å) is contained within a single "octave," leaving open the intriguing possibility that the colors we experience within the sensorium as a continuous rainbow from violet to red correlate directly to their upper harmonics or subharmonics—the higher or lower octaves—of the electromagnetic spectrum of presentation. Perhaps what are perceived as complementary or clashing colors and patterns are the counterparts of the aesthetically rational chordal structures and rhythms of music.

The evolution of the sensorium can best be conceptualized as a time-lapse sequence of multisensory moments, subjective views from within the skulls of a long succession of ancestors standing upon the same lookout point—with Mount Kilimanjaro off in the distance, let us say, its silvery glaciers advancing and retreating through time. At 24 frames per second and 18 years per generation, the 4.4 million year span that separates the earliest known hominid, *Ardipithecus ramidus*, from modern humans could be viewed in less than three hours. A marathon screening (in our submersible theater) of the 500 million year period from the earliest ocean vertebrate—an armored, jawless fish of the class *Ostracodermi*—would last nearly 2 weeks. How long the theater would remain in pitch blackness before the first shadowy images appear upon the screen is anyone's guess.

Once the dim light stabilizes, the brain begins to transform conceptually into a lush, primordial tropical rain forest within the skull—a virtual biosphere teeming with life forms. The countless animal and plant species are the miscellany of neuronal assemblies, each specializing in the production of a particular sensory effect. Pressured by the ongoing evolution of neighboring virtual biospheres, the presentation structures responsible for each neurobioluminescent effect must also evolve and adapt—generation by generation—frame by frame in our time-lapse movie—to fit ever more gracefully into the various sensory cortices, their inner ecological niches.

From under the drab canopy of the giant ferns of the early Cretaceous in due course evolved the pink and purple splashes of orchids, the bold, chartreuse camouflage of the walking leaf insects, the neon blue, green, and yellow skin of the red-eyed tree frogs, and the spectacular, variegated plumage of parrots. From under the thick skull of a dimwitted forebear in due course evolved this majestic sensorium, cloistered within whose dome of vibrant colors we stroll through the rain forest, admiring the view.

RUNNING FOR COVER

A SUMMARY OF CHAPTER EIGHT

Taking an unsentimental look at the marriage of Cupid and Psyche, we find that it is Cupid, not surprisingly, who has the upper hand. The rigorous Darwinian laws of survival of the fittest ensure that only those sensorium environments fully adapted to the continued existence of the physical organism will persist through time.

The most adaptive presentation of sense data is one that implies the existence of a centralized self. The compelling presentation of this wraparound world did not arise by accident. This mode of spherical presentation has been "purposefully designed" by evolution to promote the self-centered, self-serving, self-protective, and self-aggrandizing behaviors that guarantee the survival and proliferation of the individual and the species. The "physicality" of the sense of self is bolstered by the somatosensory cortex and the sensory homunculus that anchor us directly to reality through tactile sensations of cold and heat, pressure, pleasure, and pain.

One striking feature of the sensory homunculus is its mammoth head—or more precisely, the two halves of its enormous face. They stand bolt upright on either side of the cerebral cortex—just beneath the temples of the skull—dwarfing the rest of the homunculus body (excluding the hulking hands). Their size and situation reveal their critical function in the presentation of sensorium consciousness; they provide a tangible containment field and secure, inviolable physical parameters for the sense of "self."

The structural parallels between the sensory and motor homunculi prompt me to speculate that they may together be responsible for complementary aspects of sensory presentation. The efferent nerve impulses initiated by the large Betz cells of the motor homunculus (that cause particular muscles of the body to move) may also—through the very act of firing—"present" those muscular movements to consciousness. The visceral sense of a fully integrated presentation of the "feeling" and the "doing" homunculi may be what athletes refer to as "the zone" and what musicians call "the groove."

CHAPTER NINE

MOOD RINGS

Now that we have some understanding of how the brain creates the sensorium and how the hollow, spherical sensorium in turn creates the persistent illusion of a separately existing *self*, we may be prompted to ask, "To what end?" We can be certain that the body has not gone to all of this trouble and expense on a lark. Cupid is a harsh and covetous master and expects loyal service from his captive queen.

Just how does the body entice (or compel) this indwelling awareness to do its bidding? To address this question we need to call upon an entirely new and fast-growing branch of neurobiology. Its practitioners cannot yet even agree upon a name for their discipline, for the more they investigate, the more the systems and substances seem to expand in complexity and interconnectedness. Neuroendocrinology, neuroimmunology, and psychoneuroimmunology are three contenders, depending upon where one draws the arbitrary borders.

The linguistic confusion is not at all surprising, considering the subject matter. It is not an easy task to wrap one's head around these molecular constituents of the brain—for some do not tidily confine themselves to the brain proper. Billions of free-floating molecules are secreted directly into the bloodstream by the brain and by other ancillary glands, where they circulate throughout the body, attaching to specific receptor sites upon the cell membranes of the neurons of the central and peripheral nervous system.

The effects of these purported "molecules of emotion" are not even limited to nerve cells; they attach to and influence the behavior of all types of cells—those within vital organs, within muscles, and within many other tissues of the body. Blurring the boundaries of the traditionally discrete systems of gross anatomy—the circulatory system, the lymphatic system, the nervous system—a new and amorphous *peptidergic system* is upsetting the applecart of compartmentalized subdisciplines, blurring the conventional divisions of body and mind. The relatively recent discovery of this growing group of functional molecules has opened a Pandora's Box from which has emerged a great swarm of more than a hundred unruly *neuropeptides*.

The foremost proponent of the scientific hypothesis of a complete physiological body-mind integration is Candace Pert, author of *Molecules of Emotion: Why You Feel the Way You Feel.* She is best known for her discovery in 1972 of opiate receptors, which act as molecular locks on the surface of brain cells, activated by neurotransmitter keys—in this case naturally produced opiates called endorphins. The word endorphin is an abbreviation of "endogenous morphine," which means a morphine produced naturally within the body. Her investigations showed how heroin, codeine, and other *exogenous* opiates function by utilizing the same mechanism.

Pert's research has led her to the conclusion that a certain category of protein molecules—long strings of amino acids known as neuropeptides and neuromodulators—are responsible for establishing and modulating our emotional states. Particular neuropeptides are associated with fear, others with anger—still others bring on feelings of contentment or states of elation and bliss.

Pert's work in the early 1970s helped catalyze and legitimize the field of mind-brain-body integration, bringing it from the borderland of the holistic health movement into the halls of conventional medicine. She believes that the whole body is interwoven into a global psychoimmunoendocrine system of molecular intercommunication, linking organs and muscles, the immune system, the nervous system, and the brain. The discovery of neuropeptide receptors in all of these tissues has helped validate this claim. According to Pert, "Emotions are the nexus between mind and matter, going back and forth between the two and influencing both."

To get a sense of how neuropeptides affect life in the sensorium, consider the following thought experiment. First imagine the sensorium operating in an ideal setting. Picture yourself on a warm tropical beach, stretched out in a hammock slung between two palm trees, at the end of the first week of a month-long vacation, gazing dreamily at the puffy, white clouds floating through the cerulean sky. The waves are gently lapping the shore; the delicate sounds of a distant marimba drift in and out of your consciousness. Settling into a long and luxurious period of safety and contentment, your brain can now afford to release its store of happy neuropeptides. Soothing waves of endorphins and wisps of dopamine wash in and out over the convolutions of your cerebral cortex. All is bliss. No troubling memories of office politics, childhood traumas, or domestic intranquillity can be entertained now. They find no place to resonate within the major chords of the joyful musical metaphor presently being played by your inner marimba band.

Now consider an entirely different brain. This brain is very agitated—fearful—quite near panic. Its indwelling awareness had just done a very stupid thing. It had forgotten to stop at the filling station fifteen miles back, and the car has run out of gas on a deserted country road; and now, stumbling back toward town in the cold, pitch-black night, with no jacket and an empty gas can in your hand... (Oh yes, did I forget to mention? This is also *your* brain, a few days after returning from your tropical vacation). A burst of norepinephrine and epinephrine secreted from the adrenal medulla (located above your kidneys) is having its way with your thought patterns. Your parasympathetic nervous system has been completely overstimulated. Your heart rate has increased dramatically, your metabolic rate as well. You can feel the tightening in your chest from the increased blood pressure and vasoconstriction. Your anxiety is intense. You are ready to kick ass or hightail it out of there at a moment's notice if you have to. What's on your mind now? "How could I have been so stupid? What was I thinking? It's after midnight! Where the hell am I? What if the gas station is closed? What was that noise? Didn't I read last week that there were mountain lions on the prowl around here? Stupid! Stupid! Stupid!"

You have completely lost access to any calm, serene thoughts. The pleasant mental slideshow of last month's tropical vacation that you were enjoying just twenty minutes ago, driving home from a lovely dinner party with friends, couldn't be brought to mind now if you tried. The slight afterglow from the cognac that was warming your thoughts has entirely vanished. That brain has been completely replaced by this one. It's a new instrument; it's playing in a different key. The firing properties of all of its neurons have been adjusted by this new flood of neuropeptides. It cannot resonate the same patterns it did before.

At any given moment, the brain is in a particular state, with its own particular feeling tone. In that state it has access to certain brain functions, certain memories, and a certain range of associated feelings. For the moment, other functions, memories, and feelings are unavailable, or at least much more difficult to gain access to. As a cascade of neuropeptides floods the brain, a wave of anger or joy, fear or sadness washes over the nervous system. The brain is the tasteless tofu that soaks up the latest marinade—whatever mix of opiates and any of well over a hundred other neuropeptides: acetylcholine, serotonin, dopamine, GABA—to name but a few of the more commonly known varieties. The resonance patterns that can easily be sustained in one mood do not function properly when another tide of emotion flows over the brain.

It is a continuous feedback loop, in which the emotional coloration of the sensorium shifts in response to Psyche's most current evaluation of the state of the external world—this shift in turn coloring all subsequent perceptions. The change in mood correlates with a quantifiable alteration in the oscillation patterns of the brain. Therefore, in a manner of speaking, Psyche may be said to possess many *different* brains, each one adapted to a different set of circumstances, which she inhabits according to her *mood*: a brain of anger, a brain of sadness, a brain of joy. The stabilization of a mood—including even the subtlest, most fleeting mental or emotional predisposition—represents the establishment of one particular complex, chaotic standing waveform produced by a certain subset of neuronal resonators—those that can sustain their patterns of oscillation within the current neuropeptide environment. As the *Light* generated by these firing presentation neurons brightens or dims, the mind dutifully accommodates the shift to the new *brain state*. Whether from blissful to bland or from pleasant to nightmarish, each new neuropeptide environment causes a sea change in the prevailing emotional tide, with the appearance of the world shifting accordingly, to become either a place of safety, of danger, or of tantalizing opportunity. Like the set chord progressions of a particular musical genre, the instrumentation is familiar, the voicings predictable, and the possibilities for improvisation greatly constrained. If you are "singin' the blues" expect to hear a lot of minor thirds and flat fives.

We are all familiar with the major categories of resonance patterns, because we cycle through them every day. If a cap with a high-density array of electrodes is positioned upon the scalp, the electrical activity of the brain can be displayed on an electroencephalogram's readout in the form of oscillating brain waves. When the brain is engaged in mental activity, it generates high frequency, low amplitude *beta* waves, ranging from 15 to 40 cycles per second. The resting brain—or the brain in Zen meditation—produces lower frequency, higher amplitude *alpha* waves, from 9 to 14 cycles per second. Periods of drowsiness, daydreaming, or creative ideation show up as *theta* waves on the readout. They are even slower and of greater amplitude, from 5 to 8 cycles per second. In sleep, *delta* brain waves predominate; they are the lowest frequency and highest amplitude waves, from 1.5 to 4 cycles per second. During REM sleep, while dreaming, *theta* waves re-emerge. In deep, dreamless sleep, one drifts back into low *delta*.

An electroencephalogram's measurements of brain waves are simply too coarse to reveal much at all about the subtle microscopic resonance patterns that create the sensorium display. They do give an indication, however, of the vast and intricate nesting of systems within systems of periodic resonance that must be functioning concurrently in the brain to produce such a degree of large-scale "chaotic order." The brain is a *chaotic resonator* in the special mathematical sense of a system capable of moving between and settling into a number of complex patterns of resonance called "attractor basins," which are subjectively perceived (from within) as our various emotions, moods, and predispositions.

Chaos Theory provides the framework for a growing number of branches of nonlinear science and mathematics that study complex open systems. It has been used to analyze weather patterns, stock market fluctuations, earthquake probabilities—everything from particle physics to complex biological systems such as the brain. Chaos Theory gives us a way to talk about those systems that operate somewhere between predictability and randomness, between harmony and noise, always changing and always erratic, but kept within certain convoluted boundaries and displaying complex structures of chaotic order. This cursory introduction to Chaos Theory is only intended to provide a context for the term *attractor basin*, since we will be using this concept to help us understand the functional processes of mood and mood-shifting. We will take up the important subject of chaos again in chapter seventeen.

An attractor basin describes a mathematical pattern that regularly appears in the data from measurements of many types of brain activity—in levels of electromagnetic charge, in fluctuations of microvoltage, in firing rates of individual neurons, and in the oscillation patterns of large neuronal assemblages. An attractor basin, as a conceptual model, resembles the object from which it gets its name: a gently moving shallow dish or basin into which a marble has been dropped. There is one low point, or perhaps two or more, at the bottom of the basin to which the marble appears to be "attracted" as it moves in its irregular looping patterns. Given enough energetic input, the marble may move out of a given attractor basin and fall into another.

There is much evidence, clinical as well as anecdotal, that each particular mind state such as anger, fear, or joy operates as an independent attractor basin, promoting the recollection of its own "state-specific memories." These are the memories that were laid down during previous episodes of the same emotion. The whole emotional landscape is laid out like a topographical map of the backcountry. The emotions are the various valleys—all of which can be visited, but only one at a time—each offering its own beauties, challenges, or dangers. Some of these valleys have become very familiar to us. Since we know them well, we may choose to "hang out" in them most of the time—although they may not be the most interesting or desirable locations. If we spend our days in the Valley of Anger or the Valley of Self-Pity, we should bear in mind that the lake in which we fish for our supper will undoubtedly be contaminated with the same emotional toxins.

Each time a mind state is revisited, the associated memories reformulate and reappear in the mind, which tends to reinforce and validate the current mood in a positive feedback loop. If these thought patterns are indulged in for an extended period of time, neuromodulators and hormones in the blood may act to deepen a particular attractor basin, making it all the more difficult to exit. In chapter nineteen we will find out more about the secret trails that lead up to some breathtakingly beautiful high valleys—states of meditative equipoise and ways of being that are joyful, compassionate and loving. Countless previous sojourners have kindly carved out switchbacks up the steep, rocky slopes, and if we choose, we can follow in their footsteps to catch our first glimpse of the blissful Valley of Shangri-La.

The neuropeptides associated with a given mood affect the firing patterns of the neurons responsible for the projection of the sensorium. They do so by a variety of means, but most directly by adjusting the rates of diffusion of neurotransmitters across the synaptic cleft (the point of "contact" between neurons). Neuropeptides with either excitatory or inhibitory properties can increase or decrease a neuron's rate of firing. Additionally, the class of neuropeptides called neuromodulators, mentioned in the paragraph above, can have a more global effect upon the nervous system, causing moods to linger for minutes, hours, or days. Neuropeptides may increase or decrease the volume of activity in one or another functional region of the brain, stimulate or inhibit the growth of new axonal and dendritic connections, and deftly fine tune a thousand other sensory and cognitive features.

Such a wide variety of neuropeptides multiplies the utility of the brain a hundredfold. We should no longer view the brain as a single organ comprised of the totality of its neuronal connections—which may shed some light upon the question of why it appears to be such an inchoate clutter. This is because it is a simultaneous overlapping of a hundred different brains, only one of which is functional at a time. The brain is like a stack of transparencies—tucked into a manila folder beside the overhead projector—upon which the professor has written the charts and diagrams he will be presenting during his lecture. If he were to take the entire stack out of the folder, place the stack upon the illuminated glass, and project it all at once onto the screen, it would appear as a massive jumble of confusion—totally indecipherable. But if he places the transparencies one by one in sequence upon the glass, the students can clearly see how his thoughts are organized.

A sentient being is not just gifted with one brain; she has hundreds of models to choose from. Each stable brain state, with its own emotional coloration, allows Psyche to function properly and behave effectively in the circumstances present to her perceptions. When she is angry she has access to all of the vivid memories of other states of anger that she has experienced over the course of her lifetime. What may have started as a mere inkling of annoyance encounters a related memory and becomes a flash of anger that may be worked up into a fury through the power of concentration. In that emotional state she activates specific brain areas, archetypal memories, and an arsenal of mental weaponry to which she can only gain access when true anger is piqued. In that state of mind she is able to build an ironclad case for whatever action she feels she needs to take.

It may be a bit misleading to call these neuropeptides "the molecules of emotion," for *they* are not experienced directly. It is not as if each of these molecules were of a particular flavor, perceived by the body-mind like the taste of a spice. Oxytocin doesn't put us in the mood for sex because it smells like erotic incense; the euphoria associated with dopamine does not arise because the chemical tastes like an exquisite Cabernet. A shift in mood reflects the *global* realignment of the entire mind-brain-body complex, with the emerging emotional overtones affecting the coloration of every aspect of the sensorium experience. The sensorium operates in an entirely different manner in the presence or absence of the various neurotransmitters and hormones that modulate our emotional states. Indeed, the whole body is affected through a cascade of chemical and biophysical modifications when there is a sudden change in the prevailing mood.

The abrupt alteration of firing patterns forming the overall *standing wave* of neuronal resonance within the sensorium is the primary cause of the particular "feel" associated with a newly emerging mood. That *subjective feel* has been sculpted by evolution as meticulously as any blackbird's wing or tiger's tooth. To put it in the most rudimentary terms: good moods are intended to feel *good*; bad moods are intended to feel *bad*. This is how the body motivates consciousness to move toward the desirable and away from the aversive.

There are states of being—ways in which Psyche can be situated within her palace—that feel comfortable, alluring, or pleasurable to her. They can be broadly categorized as the "good" moods and most likely correspond to harmoniously structured, beautiful, or intriguing patterns of neural resonance. The aversive, or "bad" moods—during which the neural resonance chamber bangs out its frightening, chaotic chord progressions, all out of tune—should not be viewed as *bad moves* on the body's part (although we may wonder—unconsciously anthropomorphizing our own flesh body—why it would impose such cruel and torturous punishments upon its own *self*). On the contrary, each one of these aversive moods is painstakingly designed by evolution to be *untenable*. Each is a deliberate, *temporary state of discomfort* imposed by the physical body upon Psyche, whose job it is to extricate herself from the uncomfortable situation by addressing the underlying problem with an appropriate response: a movement toward or away from the troublesome stimulus—an attack or a retreat. For this service she will be rewarded with the neuropeptide equivalent of a warm and reassuring hug from her Skinnerian captor, Cupid.

We notice our moods most when they suddenly shift. A change in mood is usually intended to bring some new environmental or interpersonal development to our attention. Therefore, if we dwell in a mood, particularly an inappropriate mood, for too long a time, our perceptions of reality can no longer be considered completely valid or trustworthy. They will be colored—perhaps by a rosy pink glow for weeks on end when we first fall in love, and everything is delightful and serene—even that adorable mountain lion approaching us on the trail. Or we may be stuck wearing the dark blue glasses of a lengthy depression, during which a once rosy scene now appears threatening and bleak.

If a mood accurately reflects current real-life circumstances, then the survivability of the animal experiencing this adaptive mood will be greatly enhanced. The entire organism can thus be fully dedicated to the particularities of the moment. If the neuropeptides and neuromodulators in its body-mind can be skillfully readjusted in this way, the same animal can manifest as a tender and caring parent in the den and a vicious, bloodthirsty predator on the open savanna.

Once this division of labor between separate brain states had been established in some ancestral animal form, each brain state would find itself subject to *independent* circumstantial selective pressures on its evolutionary success. Here I mean to draw attention to the curious fact that brain states evolve largely independently of one another, in the same sense that an elephant's ears and tusks evolve "independently." Of course the ears and tusks are attached to the same animal and must work in coordination with one another, but they each *evolve* in response to different evolutionary pressures that are expressed circumstantially at completely different moments in time. For natural selection actually to produce such large ears, which function for the elephant primarily as cooling thermal radiators, conditions must occasionally become so extreme that they actually affect the elephant's survivability, and we can imagine that this only occurs on very hot days during very long droughts. The pressure for having long and sturdy, well-mounted tusks is only put to the acid test during life-and-death battle, when the elephant's very survival is at stake (ignoring for the moment the important matter of sexual selection and the conferral of reproductively advantageous alpha dominance status as a cofactor affecting the evolution of both of these particular features.)

A certain hypothetical species of animal may be very effective at responding to dangerous circumstances, having fine tuned its fight or flight response so as to flee at the very first sign of danger—perhaps because it is unlucky enough to inhabit a niche where it is on the menus of several disparate predators. Its extremely alert "norepinephrine-epinephrine-acetylcholine brain" may be very well adapted for cautiously scampering about in search of seeds and nuts in the bright sunshine. Later on, having retreated to the relative safety of its burrow, another concoction of neuropeptides and a completely different mood may now be indulged in, perhaps one of playful curiosity or restful relaxation—such moods having proven adaptive in this more secure environment.

Suppose that this hypothetical animal—due to some whimsy of its evolutionary history—must also court and mate aboveground, where at any moment it may as likely encounter a predator as it may a sexual partner. It would have been forced to evolve a hair trigger mood-switching mechanism between flight and sexual arousal—*its* survival quandary now being much more vexing: *"Mate or be eaten?"*

Those moods that affect survival most directly in a given species become more and more idiosyncratic, for they must be independently adapted to address effectively each of the prototypical scenarios that comprise the organism's daily, monthly, or annual routine. Transitions, in general, should be relatively abrupt; Psyche must remain ever vigilant and agile, ready at any moment to "incarnate" into whatever newly created emotional "being" is next up on the roster.

This division of the brain into discrete *states*—called *emotions, moods,* or *predispositions*—gives evolution exactly the kind of raw material it needs to fashion the specific "emotional survival kit" most appropriate for each new species of animal, as it emerges to exploit some new niche in the ever-changing environment. On an evolutionary time scale, stable moods can shift and merge, differentiate and dissipate according to the needs of the evolving new species. The precise palette of emotions appropriate to the African wildcat, *felis silvestris lybica*, may not work so well for its evolutionary successor, the domesticated housecat. Any serious atavistic tendencies suddenly re-emerging in the body-mind of our darling Fluffy may yield her an unanticipated and most tragic trip to the pound.

Each mood has its own flavor, its own feel, for at least two very good reasons. First, this allows Psyche to learn to *identify* the moods she is entering, and having gained experience in how certain moods succeed or fail in particular situations, she is better able to steer her emotional craft. The second adaptive use for having a full spectrum of identifiable functional moods is that consciousness is thereby able to conjure up an emotion out of next to nothing. By "attending" to the feel and flavor of a nascent *proto-mood* as it is called up from memory, Psyche is often able to bring on a full-fledged emotional state at will. Her judgment of the utility of a particular emotion is not, as in the first example, *caused* by immediate sense data from the outside world but rather comes from a calculation, intuition, or memory of a previous situation that leads her to conclude that the present circumstances would best be met with that particular emotional response.

Now that we have some sense of the type of marriage that Psyche and Cupid have entered into—forming the sentient being—and we understand better the roles of the two parties concerned, we will next look through the scrapbook of their early courtship. The "photos" are fossils; therefore, the images are indistinct. In the next two chapters we will look back in our imaginations to the very first appearance of conscious awareness in the earliest true animal forms (animal: from the Latin *anima*—breath, soul). How long she lived as a dreamy blur, a sometime companion to the body, phasing in and out of physical reality, we cannot know. What tempting features first lured her in? We will try to imagine Psyche in her earliest moments of physical existence, at her first conception of the elemental thought—*"I am..."*

MOOD RINGS

A SUMMARY OF CHAPTER NINE

Psyche, in whichever of the myriad animal forms she incarnates, inevitably ends up spending most of her day hovering in the neutral zone between pleasure and pain—for Cupid must be able to utilize the two prodding tools of desire and aversion to get her to do his bidding—prompting her to move toward or away from other objects, beings, and environments in a perfectly adaptive manner, one that serves to maximize the survivability of that particular organism.

Through the judicious release of well over a hundred neuropeptides, neuromodulators, and hormones, the brain is able to create at will a number of divergent brain states, each one adapted to a specific set of circumstances that the organism is likely to find itself in at any given moment. These distinctive brain states are experienced by consciousness as its various moods, emotions, and transitory predispositions. Each newly emerging mood or emotion corresponds to a global reconfiguration of neuronal firing patterns within the brain—a mechanism intended to refocus consciousness quickly toward the particular issues at hand. The full complement of moods and emotions has been carefully crafted by evolution to permit the organism to respond adaptively in a wide range of real life situations.

Perhaps the most difficult thing for us to see and accept in this whole theoretical construct is just how ruthless the body can be in relation to its own indwelling consciousness. We might prefer to remain in blissful ignorance, rather than face this awkward truth and risk a moment of existential anxiety—this emotional reaction itself being one of Cupid's stinging darts. However we may sugarcoat it, the fact of his treachery is undeniable—just consider the body's callous presentation of painful sensations and aversive emotional states. The body pleasures us only "at its pleasure" and punishes us just as readily whenever it sees fit.

CHAPTER TEN

THE THREE POISONS

Buddhism, I think it is fair to say, has a reputation for presenting a rather dour and pessimistic view of life. *Buddhists,* on the other hand, whether they are Dharma teachers or longtime practitioners, seem to me to be among the happiest, most open and confident people I have ever met. There is something therapeutic about a clear-eyed admission of just how bad things really are that allows one to face the doubts and fears that otherwise lurk in the shadows of the world of a cockeyed optimist.

Buddhism is the study of what is under the surfaces of things. It looks at the world and sees—beneath the self-evident beauty, the glorious wonder, the sensual pleasure and the raucous good times—something disturbing that it feels compelled to investigate. These inquiries led the Buddha to discover the three irreducible obstacles to finding a permanent state of happiness in any particular configuration of physical or mental constructs. They were referred to in chapter five as the Three Marks of Existence: *emptiness of self nature, impermanence,* and *unsatisfactoriness.* This is a way of talking about the problem circumstantially—acknowledging the futility of the search for happiness in the rearrangement of the specific conditions of one's life,

THE THREE POISONS

since any such novel arrangement will likewise be impermanent, unsatisfactory, and empty.

It is perhaps easiest to see the root causes of unhappiness in this way—as being elemental to the universe—but this is not technically correct. *Nirvana* and *samsara*, liberation and bondage, always permeate each moment; they represent two potential experiences of the same circumstance. There is nothing about this particular moment of existence that prevents you from immediately awakening to a state of bliss—or happiness or contentment or freedom, as you prefer.

Another Buddhist analysis focuses instead upon the sentient being, to try to understand how a second triple threat—the Three Poisons of greed, hate, and delusion—can also be viewed as the root causes of unhappiness, representing the three fundamental characteristics of the animate form. Although this analysis is incredibly insightful, I also believe that this is where early Buddhist philosophers attempted to grasp something just beyond their reach; yet this is where a clear and precise parsing of reality is of utmost importance. There seems to be a mixture of baby and bathwater here. To decide what to throw out and what to embrace, we will need at least a basic understanding of genetics and evolutionary biology and a smattering of cybernetics (the study of information processing, feedback, and control in biological and other systems)—seasoned, as always, with a *soupçon* of numinous imagination.

In this chapter, we will review the basic mechanisms that have allowed complex animals to evolve on this planet. In the next chapter I hope to show how these same mechanisms are likewise responsible for the smooth transition between early cybernetic life forms, which I refer to as automatons, and life forms such as ourselves that operate by means of conscious awareness within a sensorium.

The Buddhist analysis is so insightful, so daring and so colorful that it still provides the perfect starting point for this investigation. We will begin with a taste of the three poisons.

> *All my ancient twisted karma,*
> *From beginningless greed, hate, and delusion,*
> *Borne through body, speech, and mind,*
> *I now fully avow.*

This is known as the Repentance Prayer, and it is chanted regularly in the Soto Zen tradition. In essence it is a simple and straightforward acknowledgment of one's shady past. Having incarnated countless times, I presume I have attacked and killed many beings, stolen their food, raped and pillaged, lied and cheated, and caused much mayhem. Within the four lines of the prayer, there is no explicit mention of remorse, no specific penance demanded, and no stipulation to do no further evil. This is intended to be a moment of clear seeing into the way things are and always have been. It is a recognition of our own participation in the wheel of life and death—a wheel that has been spinning for a very long time.

In Tibetan Buddhist iconography, the three poisons (of greed, hate, and delusion) are symbolized by three animals (the rooster, the serpent, and the pig), who chase one another around and around in an endless ring, forming the hub of the *bhavachakra*, or the Wheel of Becoming. This wheel is richly illustrated in Tibetan *thangka* paintings and functions as a kind of mnemonic map of the entirety of the Buddhist teachings on the topic of *samsara*, or cyclical existence. Surrounding the three animals in the hub—and due to their ceaseless activity—arise the six realms of being, separated by the spokes of the wheel: two heavenly realms (of pleasure and self-righteous striving), two hell realms (of craving and suffering), the animal realm (of unreason and desire) and the human realm (of constant change). These may be viewed either as actually existing worlds or as the various impermanent psychological states one enters throughout the day. Around the rim of the wheel is a pictorial representation of the Twelvefold Chain of Causation, the complex of forces that propels the mindstream through endless rounds of birth and death, with the mechanism kept in motion by the churning activity at the hub. The wheel is held by Yama, the fearsome Lord of Death.

It all looks rather bleak. However, it is considered most auspicious that we reside in the human realm, for it is here, with the perfect combination of pleasure, suffering, and leisure time for contemplation of the Dharma, that we may well awaken to our true nature and get off the wheel for good. Alternatively, if we hold to the view of the Mahayana schools, this may be the moment we finally join up to offer our services in the collective project of all of the countless Buddhas and Bodhisattvas: to end the inexhaustible delusions that keep the wheel spinning; to learn all of the innumerable teachings; and to help each and every one of the infinite number of sentient beings to awaken. A formidable task, but with excellent long-term job security!

What we will be investigating in this chapter is an aspect of the "... *ancient twisted karma... borne through body...*" which, as I mentioned in chapter five, I do not believe to be karma at all. The Buddhist tradition was unaware, as it formulated its cosmological philosophy more than two thousand years ago, that there could be an actual physical mechanism lurking within animals—invisible to the naked eye (and only indirectly visible to today's most advanced scanning tunneling microscopes)—a mechanism capable of secretly encoding directions for passing on fangs and claws and gaudy tail feathers, musk and venom glands, and even violent and lustful behavioral tendencies from one generation to the next. This mechanism not only re-creates these bodily manifestations of the three poisons in subsequent generations, but was also directly responsible for creating them in the first place.

Deoxyribonucleic acid, or DNA, is of course that mechanism. With the discovery of its double helix structure in 1953, evolution by natural selection was finally vindicated as a fully persuasive and comprehensive theory of the origin of life on earth. DNA is bound in structures called chromosomes, resembling twisted ladders whose spiraling sides function like twin backbones that allow the two strands to be unwound and separated during mitosis. Each strand becomes a template for two new daughter strands of DNA that segregate into the two newly formed daughter cells. The helical sides are merely structural elements, made of repeated sets of sugar and phosphate molecules; the genetic information is held in the rungs of the ladder. Along the length of either spiral, a sequence of nitrogenous bases with the initials A, T, C, and G can be read. Long strings of these letters are used to name the discrete DNA sequence, called a *gene*, that encodes a particular protein. Genes are separated by "punctuation marks" called start and stop codons.

DNA has the five basic characteristics necessary to produce complex life forms. It can store information, encoded in the alphabet of matched base pairs of adenine-thymine and cytosine-guanine. It is self-replicating, which means it can produce identical copies of itself. It can make various forms of ribonucleic acid, RNA, which in turn transforms the information stored in the DNA into the biological machinery of life—by creating the wide variety of proteins necessary to build and maintain the machinery. It orchestrates the one absolutely crucial biological function of reproduction, in which the DNA instruction manual is passed on to the next generation. Finally, for evolution to occur, DNA must be subject to random mutations in its code; it is of critical importance that it "messes up" occasionally, but only very occasionally.

Random mutations in the DNA molecule produce variations in the population of a species. The exigencies of life prune back the less fit, resulting in a tendency toward improved adaptation from generation to generation. Body structures become more organized and efficient, and behavioral strategies become more effective. In this way the general trend toward greater complexity in the service of these behavioral and bodily adaptations eventuates in the creation of the fantastic array of animal forms that now grace our planet.

This view of evolution as a life-and-death struggle for the survival of the fittest species or the fittest individual in fact represents only the first approximations of something even more menacing going on beneath the surface. If you really want to deconstruct your sense of *self*, I suggest reading Richard Dawkins's *The Selfish Gene*. When examined more closely, all of our most cherished sentimental notions of altruism, of maternal care, of mating pair loyalty—even the original cause of the division of the sexes—all devolve into a precise mathematical calculus of advantageous replication rates of favored genes. The competition between two versions of the same gene, called *alleles*, is as fierce at the molecular level as the competition between two bighorn rams in rutting season.

Since the moment Psyche first took refuge in physical form, these same unyielding forces of evolution have been working their wily ways to manipulate her experience of life in the sensorium. These "selfish genes" that produce the body and the structural framework of the mind don't care a whit for *her*. They don't have anything to care with, being truly mindless machines. *She* is the awareness, the body's only conscious experiencer of such emotional states. The palace that Cupid has prepared for Psyche can be filled with shimmering beauty in all six sense realms. This is its great allure. There is only one drawback—Psyche is not free to go when times get tough. One moment she is drifting into a pleasant reverie while picking berries by a stream, the next moment a charging predator suddenly appears in her peripheral vision, and she is pulled back to command central like a marionette and strapped into her unenviable throne, to bark orders at her hastily assembled generals and ministers of state. Psyche is routinely tied down in this way by countless Lilliputian ropes, "chemical bonds" with cascading emotional effects designed by her own genetic host. She is subject to ten thousand irresistible snares perfected over hundreds of millions of years of brutal evolution. Psyche must serve the body by playing the role of Ego.

Perhaps we should take a breather, to reflect for a moment on what all of this means to us personally. If this is the first time you have considered the relationship of your mind, your body, and your ego in such harsh, unsentimental terms, you may find yourself experiencing a strong aversion to these ideas. There is no cause to be dispirited. These are just the facts of life. As they say in Alcoholics Anonymous, the first step is to admit we have a problem. Think of this as being like an intervention. We are, each one of us, addicted to the ego, and it's important that we come to terms with the fact.

Drowning in a great ocean of insecurity, we cling to the ego as if to a piece of floating debris, one of those compulsively rearranged deck chairs, perhaps (from that infamous metaphor of futility), as we struggle to escape from the undertow produced by an eternally sinking *Titanic*. "Let go!" every God-intoxicated mystic has said. "Don't cling!" is every Buddhist sutra's refrain. "Follow the I-thought back to its source," whispers Ramana from the abode of nonduality. "Follow Me," says Jesus, with a motioning gesture toward his heart.

In the words of Zen Master Dogen (1200 to 1253):

> *To study the Way is to study the self.*
> *To study the self is to forget the self.*
> *To forget the self is to be enlightened by all things.*
> *To be enlightened by all things is to remove the barriers*
>   *between one's self and others.*

In chapter nineteen we will take a more in-depth look at the spiritual practices and psychological insights that can reengage the body-mind in a brand new, healthy, and open-eyed partnership—an integration that is relaxed, peaceful, and joyous. In a sense this is not a practice at all, but a letting go of all practices, a return to pure awareness. It certainly need not be a battle or even a struggle. What it takes is a fine discernment, a careful listening to our own thoughts and feelings, and a willingness to challenge the selfish ones with a gentle but persistent "Says who?"

The Three Poisons are customarily presented as broad categories of negative, emotion-driven, and highly immoral thought and conduct, under the formidable headings of Greed, Hate, and Delusion. The equivalent terms *desire, aversion,* and *ignorance* are also used in this context and are much better suited for our purposes—being somewhat less judgmental and more descriptive of the basic animal behaviors to which they ultimately refer. Ignorance, here, is not stupidity; ignorance is simply *indifference*—the ability of an organism wisely to ignore that which is neither a source of food or sexual gratification nor a dangerous mortal enemy. The two basic postures of reactivity, the positive and the negative, branch off in opposite directions from the zero point that represents an even more fundamental, energy-efficient nonreactivity. Altogether, these three stances of desire, aversion, and indifference provide a primitive, yet completely adaptive behavioral repertoire for our generic organism. This unholy trinity is *poisonous* only to the spiritual aspirant who seeks no longer to be bound to this mortal coil.

And now, back to our story. The evolution of animal life on this planet can be divided into several distinct eras. The earliest was the *prebiotic* or chemical era, in which the first complex molecules formed from the random collisions of simpler molecules and atoms in a lifeless sea—with additional boosts of energy provided by frequent lightning strikes and solar radiation. Macromolecules arose that were able to self-replicate by providing a blueprint matrix upon which a number of simpler constituent molecules could imprint, thus forming another macromolecule, identical to the first. With the arrival of these earliest self-replicating molecules, the laws of evolution first came into play. Coordinated systems of macromolecules with greater reproductive vigor proliferated in the early oceans.

Recent research suggests that of the three crucial cofactors for modern replication (proteins, RNA, and DNA), RNA was the first to form, though it is unclear how it was able to self-replicate. Once the

THE THREE POISONS

final tripartite format was established, selective pressures favored the ancestral DNA systems that colonized for mutual defense, and the inherent logic of cooperation, organization and specialization eventually led to the creation of the first cells and multicelled bodies.

Energy is required for biosynthesis, and a division eventually occurred between systems that utilize solar energy directly through photosynthesis (plants), those that consume live plants (herbivorous animals), those that consume decaying plants and animals (scavengers), and the real troublemakers, the ones without which the three poisons of greed, hate, and delusion could never have gotten a foothold, the *carnivores*. They are the predators that chase after the herbivores, making the food chain suddenly a much more dynamic and dangerous place. They are the ones that cause the hierarchies of fear and aggression, with small predators subject in turn to predation by larger, more aggressive animals. (All the while, swimming silently through the inner oceans of these animals' bloodstreams, are the countless bacteria and viruses, some of which will likely have the ultimate or should I say *penultimate* word, taking down the largest, most formidable predator with some devastating disease, before the fungi and mold finally arrive on the scene, to turn everything back into soil.)

This division of roles characterizes the phase of evolution in which animals develop various modes of propulsion to engage in purposeful movement—toward food sources, safe havens, and potential mates and away from predators, life-threatening situations, and disagreeable environments. Archaic sense organs first appeared some 600 million years ago with the great evolutionary leap from sponges to sea jellies. Initially the sense organs were directly connected to the motor apparatus by a primitive nervous system. The first neurons were little more than impulse conduits; the sense organs themselves accomplished the necessary information processing. These early sense organs were capable of differentiating only a few stereotypical scenarios, and the responses they generated were simple involuntary reflexes.

As sense organs became more complex and began to distinguish more detailed information about the surrounding environment and as the means of locomotion became more refined, an intermediate organ appeared—made up of much more complex arrangements of nerve cells—that took on the function of processing the datastream received from the sense organs before initiating a second stream of impulses directing the movement of limbs. This was the crucial turning point—the advent of the primitive brain. It marked the beginning of the *cybernetic era.*

Is it possible to conceive of a planet upon which there was a great deal of running and swimming about, chasing and being chased, frightful jaws of predators gnawing through the flesh of struggling prey, but without the vaguest hint of suffering? Can a complex world of predator-prey relationships—showing evidence of elaborate behaviors, strategizing, learning, and novelty—exist without consciousness? If so, is there a credible means of transition from that nonconscious, cybernetic world to the one we now inhabit, in which consciousness, at least in ourselves, is a self-evident fact?

If we are conscious now, at point C, and are living in an utterly convincing sensorium with six interwoven sense fields, subject to pleasure and pain, and we can imagine a point A at which sensorium consciousness did not exist—but fully functional sense organs and a brain and a nervous system and complex behaviors did—how do we account for that transitional point B, in which sensorium consciousness first appears in the organism?

These are obviously a lot of good questions, but the question is: are there any good answers?

THE THREE POISONS

A SUMMARY OF CHAPTER TEN

This chapter begins with a discussion of the Buddhist concept of the three poisons: greed, hate, and delusion, offering them as an appropriate philosophical framework for discussing the forces that guide the evolution of all forms of animal life.

The purpose of insisting upon the dichotomy of body and spirit is to highlight a dreadfully real and most unfortunate dynamic occurring between these two interwoven yet irreducibly distinct realms of existence. It is essential that we come to grips with the fact that many of the evolutionary adaptations that are clearly advantageous to the physical organism (or, more properly, to its genes) may nonetheless represent disastrous setbacks to what we can only hope might be some larger, divine or cosmic program of spiritual evolution. The insentient, mechanical body uses the raw material of its indwelling consciousness for its own ends and may, if it proves adaptive to do so, freely and without compunction impinge upon the serenity and happiness of the sentient portion of the sentient being.

Predator-prey relationships are responsible for setting in motion a physical and behavioral arms race that has resulted in a planet dominated by the afflictive emotions of fear, hunger, aggression, and lust. These unfortunate yet unavoidable evolutionary trends have encouraged and rewarded, through a hopelessly biased genetic lottery, a multitude of selfish and spiritually problematic behaviors.

CHAPTER ELEVEN

MISTAKING THE MAP FOR THE TERRITORY

Darwin's theory of evolution by natural selection seems, on the face of it, so simple and self-evident that one wonders how its mechanism could have remained a mystery for so long. The theory itself has evolved considerably since it first sprang from the forehead of Charles Darwin in 1859. *Neo-Darwinism* (also referred to as the Modern Synthesis) represents the merger of classical Darwinism and modern genetic theory. Now, with a much more sophisticated understanding of the physical and biochemical mechanisms that allow DNA to produce heritable phenotypic variations in individuals, with computers capable of performing rigorous statistical analyses of population genetics, and with a much more complete fossil record of intermediate forms as supporting evidence, Darwin's original hypothesis—that natural selection provides the sole and sufficient, clear causal explanation for the origin and evolution of species—has proven itself outstandingly fit. It has survived for a century and a half, successfully adapting itself to any new scientific findings—becoming stronger and more robust than ever. Well-prepared to meet any conceivable challenge, the theory of natural selection now stands firmly established as the vanguard of Rational Biology.

Though a child can understand the theory's basic mechanism, a full-grown adult can resist it to the death. No amount of careful instruction would ever be sufficient to disprove creationism to a mind not predisposed to scientific reason. For if God is indeed omniscient, omnipotent, and omnipresent, he certainly would have been aware of our inquisitive nature and the extent to which technology would advance the limits of human perception. He would have known about the future development of supercomputers, the Hubble Space Telescope, and carbon dating. He may well have buried all of those so-called dinosaur bones Himself on or about October 23, 4004 B.C.E. (the date of Creation, according to the calculations of Irish Archbishop James Ussher) as a test of our faith and likewise been responsible for manipulating the Hubble and WMAP data streams that purport to "prove" the existence of some 125 billion galaxies in a universe 13.7 billion years old.

Personally, I must confess, or profess, or, to put it more neutrally, *say* that I believe the universe is fully infused with awareness, consciousness, and spirit. I do not discount the possibility of a great deal of intervention emanating from various disembodied saints and angels, bodhisattvas and helpful deities of all persuasions. I don't think their concern is for charting the course of the evolution of the human race by picking off weaklings and ne'er-do-wells, but I could see them interceding to help us individually in matters important to our spiritual education. Maybe we each owe our very survival to the ministrations of an army of overworked guardian angels, who just want us to stick around long enough to learn something from our accumulated life experience.

I believe the Plant Kingdom possesses noble consciousness; it is brimming with life and wisdom and awareness. I believe that the earth, in a very real sense, is one mammoth creature—called *Gaia*—and that life is a mystery and a wonder. Even if none of this turns out to be true after all—for me, it is better to *behave* as though it were. Practicing kindness and respect toward all potential life forms can only be a good thing... because we never really know... and in this way the habit of loving and welcoming all beings becomes our ready response if we should ever meet up with what we deign to consider an "actual" sentient being. Regardless of whether I feel myself to be constrained by scientific reductionist reasoning in my personal beliefs, I do hope that I can play the part of a reductionist well enough here to persuade you that the advent of sensorium consciousness *can* be explained scientifically. The present theory does not postulate any divine intervention, save for that quantum spark of "divinity"—the bare awareness inherent in the quantum field itself—that ultimately comes to dwell within the sensorium. I will leave it to other brave souls to take a stab at a theoretical model for the arising of *that*.

The origin of consciousness (or the soul) is the one conundrum in all of biology that pits neo-Darwinian evolutionary theory against biblical creationism in a sparring match where, to be fair, both sides may be said to be on an equal footing. Since we humans are the only ones on the planet who *talk* about our consciousness, the notion that we are the only ones *endowed* with consciousness is difficult to disprove.

On the other hand, when biblical creationists (or perhaps I should say: proponents of "The Theory of Intelligent Design") attempt to refute the Darwinian view of the evolution of complex *physical* features of animals—the bat's wing or the human eye, for example—their arguments are invariably weak and seemingly disingenuous. They include the infamous "Watchmaker and the Watch": that the existence of a complex mechanism such as a watch implies the existence of a watchmaker in the same way that the existence of a complex organism such as a man implies the existence of a man-maker... God. They focus their attacks upon a "straw man," insisting that Darwinism posits the creation of animals by *pure chance*—comparing the probability of the evolution of man to the likelihood of an immortal monkey continuously typing random nonsense coming up with the complete works of William Shakespeare. The equally risible "Of what use would half an eye be?" is intended to make short work of the Darwinian theory of the gradual evolution of complex organs through a series of random mutations (although image-forming eyes of many types have evolved—or been "created"—more than 40 times independently in a wide variety of animal groups). The simple eye-spots of the flatworm, though possessed of only a few of the parts of the complex eyes of its more highly evolved distant cousins, can still distinguish light from dark—important clues for survival, in a pinch. In the land of the blind, the one-eyed man is king. Creationists quote Darwin himself when he bemoans the lack of transitional fossils, failing to acknowledge the advances that paleontology has made in the century and a half since the publication of *The Origin of Species* and the subsequent documentation of thousands of series of transitional fossil forms.

In attempting to conceptualize how consciousness itself arose from nonconsciousness, however, we are presented with a much thornier problem. "Of what use would half a consciousness (or half a soul) be?" is certainly a much more provocative question. It indicates the presence of a true dilemma, impossible to reduce to constituent parts or to compare with the evolution of physical structures. Even if we replace the words "half a consciousness" with "cloudy consciousness" or "intermittent consciousness"—for which an evolutionary advantage could undoubtedly be argued—it still begs the question. We cannot wish away the striking difference in the before and after pictures between an early cybernetic, robotic animal operating without even a hint of awareness and one, such as ourselves, equipped with fully realized sensorium consciousness. What we have now in our minds is absolutely unique and uncompromisingly real. For each one of us, our own subjective I-sense is an ontologically irreducible, objective reality.

In *The Emergence of Everything: How the World Became Complex*, Herald J. Morowitz describes a series of thirty-six "emergences," radical shifts in the organization of matter and energy that bootstrap our universe—latent since the moment of the big bang—into being. He traces the evolution of stars, planets, and terrestrial life—from prebiotic macromolecules to single-celled organisms, from early animals to vertebrates, hominids to *Homo sapiens*, on up through a progression of ever more complex human societies and technologies. He catalogues each of the key physical, chemical, biological, and cultural developments that are the strict prerequisites to all subsequent stages of complexification. He writes of the emergence of consciousness:

> *An emergent property related to cephalization [brain complexification] is cognition or consciousness that is the animal mind. Some biologists believe it appears at an even more primitive level. Indeed some would trace awareness back to the preneuronal forms or even to protoctistan [single celled animals].*

When scientists speak loosely of consciousness being an "emergent phenomenon"—something that arises spontaneously when cybernetic systems achieve a certain threshold level of complexity—I am left with the impression that they are using the word *emergent* as if the word itself were somehow an explanation. At best, it provides only a weak *argument by analogy* to those other essential, but much less momentous chemical, physical, and biological emergences of novel hierarchies of complexity. To my mind, this "explanation" for the emergence of consciousness is not a great deal more useful than the explanation given in the Middle Ages for the existence of rats—that they were the result of "spontaneous generation" from piles of old rags!

We should start by redefining this vague and troublesome term *consciousness* more carefully, by limiting its scope to the specific form of consciousness we each know actually exists: *sensorium consciousness*—the subjectively experienced, luminous world-surround-display. We would then have, by any definition, a legitimate "emergent phenomenon" on our hands and one whose *means* of emergence into our animal lineage may in fact be surprisingly easy to explain. In order to comprehend fully how sensorium consciousness may have first appeared within the cybernetic (information processing) brain of one of our prehuman ancestors, we will need to reexamine whether our current *computational* models really offer the best description of the functioning of an actual modern brain.

Are all neurons fundamentally on/off, if/then, yes/no logic gates? Is it helpful to view all neurons as simply the "wetware" components of a massive biological supercomputer? Are neurons really best described as information processing devices? I propose that they are not, at least not *all* of them. I believe that our bias toward viewing neurons in this way stems from the simultaneous and somewhat incestuous *emergence* during the last half of the 20th century of two eager, young, upstart fields of study that would have done well to keep their distance from one another until they matured a bit more. I'm referring to the departments of information processing, computer science, cybernetics, and systems theory on the one hand and the departments of neurobiology, neurology, neuroendocrinology, and microneuroanatomy on the other. It is clearly the case that whole regions of the brain actually *are* devoted to nonconscious information processing of one kind or another, and the more primitive the animal being studied, the more valid it is to compare its entire neuronal structure with computer circuitry.

Early success can spell death to deeper investigation, however, and as long as a scientist "knows" what neurons are, he may never discover that they actually lead a double life. Somehow, because of or in spite of all of this mindless computation, new and luminous configurations of those selfsame neurons are now responsible for the *production of* and the *presentation to* sensorium consciousness. What possible functional arrangement of neurons in the brain could have allowed for this seemingly miraculous emergence of a conscious life form from a nonconscious forebear?

In order for us to begin to grasp the magnitude of this moment of transition from presensorium cybernetics to sensorium consciousness, let me propose the following thought experiment. Imagine yourself to be an ancient creature, the very first conscious being on the planet. For reasons that will soon become clear, I cannot offer any definitive guidance as to when in prehistory this creature might have emerged. Choose an individual from a species that you yourself could imagine inhabiting—perhaps an early hominid, an ancient mammal, a reptile, or a fish—as you prefer. Don't choose Tyrannosaurus Rex, although I know it's tempting. He is not in our direct lineage and is too high up on the food chain for our purposes. The creature should be a midrange carnivore, a predator from the point of view of some animals and potential prey to others. This will ensure that it possesses the alacrity of mind necessary for this thought experiment—an intelligent responsiveness sculpted by evolution from the need to distinguish when to give chase and when to flee.

The animal that you have chosen to inhabit has already evolved a large number of discrete brain states that it is able to shuttle between through the release of various hormones and neuropeptides. Each brain state is marked by a change in the firing rates of neurons and a reorganization of global firing patterns. The animal is therefore capable of exhibiting complex and situationally appropriate behaviors. The organism has highly evolved legs or fins to move about with and complex sense organs to scan the environment.

Now ask yourself: what is it about this creature—specifically what is it about its brain—that has allowed *you*, the first conscious being on the planet, to experience your existence from *its* point of view, whereas your siblings, identical to you in every respect, raised by the same ostensibly attentive parental units, are still (as are they) nonconscious automatons? I pose this question in such a stark manner in order to point out that there clearly must have been a moment of metamorphosis, when the cybernetic brain activity of a nonconscious ancestor transitioned into the very earliest manifestation of sensorium consciousness. This radical shift in the functionality of existent neuronal structures propelled the brain into an entirely new evolutionary trajectory—in which it was required to *present* an accurate but nonetheless *illusory* portrayal of the external world to an indwelling conscious *self*.

Even if this transitional "moment" was not instantaneous, even if it took a million years to become fully established, the fact that the transition could take place at all clearly implies that there must have existed a common neuronal structure that served both the conscious and the nonconscious versions of the creature well enough. If we are not to fall into the trap of imagining that the structures that *permitted* sensorium consciousness were somehow *prefigured* for that eventual purpose, we must explain them as logical developments of a particular kind of nonconscious cybernetic organization in the automaton. This in turn will give us a clearer indication of how the sensory cortices of early animals must have been arranged and how they functioned. The evolution of those features must be impeccably logical in the *absence* of consciousness and yet be a hospitable matrix *for* consciousness. I believe that the transition could have been fairly rapid, even instantaneous, because of how the sensory cortices of early cybernetic animals were arranged. We will see how the bioluminescence of firing neurons—which was truly an insignificant epiphenomenon of cybernetic computational processes—provided the perfect "virtual environment" for enticing and sustaining conscious awareness.

What I find so attractive about this particular portion of the larger theoretical construct—which I will attempt to clarify on the following pages—is that it, in and of itself, does not require the bottom up theory of microgenesis discussed earlier (in which thought-moments appear in quick succession, originating as impulses from the brainstem), although both theoretical frames are in complete accord. It does not exclude the possibility of consciousness "descending" into physical existence from somewhere on high. At this point, there is still plenty of ontological and theological wiggle room.

One crucial cytoarchitectural feature must be well-established in the preconscious cybernetic brain in order for our hypothetical animal to be able to transition directly to sensorium consciousness: cortical structures arranged into *topographic sensory maps*. A topographic map is a discrete section of brain tissue that exhibits an organized, point-by-point correspondence to specific locations in the visual, auditory, or somatosensory field or represents some other incremental variable of sense data by means of a *spatially structured neuronal continuum*. In many cases these sensory maps are completely analogous to street maps—relative proximity and direction on the cortical map indicating relative proximity and direction in exterior space. In the modern brain, many such maps are known to exist—representing all aspects of the sensory world. Maps throughout the sensory cortices are also connected with subcortical areas that themselves reflect aspects of the original mapping.

The primary visual cortex (or *striate cortex*), located in the occipital lobe, is the largest *retinotopic* map and would seem to be a perfect candidate for the initial transitional structure. When light reflected from an object strikes the retina, which is essentially an incurvate two-dimensional sheet, the visual form of the object is then mapped directly onto the surface of the striate cortex in a simple, point-by-point topological translation. Although there is some deformation of the striate cortex image to accommodate a more detailed view of the center of focus (similar to the "exaggerations" we found in parts of the somatosensory homunculus), this straightforward, mechanical translation from the outer world to the retina to the striate cortex represents an elementary exercise in basic cartography. A dozen other retinotopic maps—for spatial orientation, color differentiation, luminosity, edge recognition, movement, stereopsis (binocular vision), and object tracking—overlay the visual cortices and associated nuclei.

Some maps are more abstract—for instance those that categorize auditory pitch relationships. Three separate sound frequency maps are found in each cochlear nucleus. Within the auditory cortex, one dimension of the *tonotopic* map is laid out just like a piano keyboard. The second dimension, it is presumed, maps an as yet undetermined correlated feature, perhaps timbre or chordal overtone structures. Accurate spatial sound maps, created through meticulous comparison and contrast of binaural streams of auditory input, collude with other maps that distinguish pitch, harmonic relationships, and tone color to reproduce in fine detail the entire auditory landscape.

A variety of topographic somatosensory maps tracking heat and cold, pleasure and pain, pressure, body position, and balance, project upon the somatosensory homunculus and the somatosensory association cortex. Altogether, five separate somatosensory cortices have been discovered, each with its own distinctive somatotopic representation, exhibiting complex cytoarchitecture and rich interconnectivity.

The prevalence of these maps should come as no surprise, even within the brain of a robotic animal completely devoid of consciousness. One-to-one correspondences between the sense fields, sense organs, and sensory cortices derive principally from the innate logic of embryonic growth patterns. The simple geometrical and spatial constraints of cell replication, as sense organs and cortical tissues develop and differentiate both ontogenetically and phylogenetically—following the line of least resistance—would clearly tend to result in the formation of logically organized maps. Natural selection would discourage the brain from making randomly *dis*organized maps—a muddled map in almost all instances being much less useful than an orderly one.

In a cybernetic animal's primary visual cortex, for instance, evolutionary pressures would for a variety of reasons favor the development of a two-dimensional, retinotopic cortical map corresponding to the incurvate two-dimensional surface of the retina. Having those neurons that are responsible for analyzing data from adjacent areas of the retina (representing adjacent areas in visual space) in close proximity to one another would facilitate a number of important cybernetic processes. It would allow functionally related neurons to relay information to one another more easily—to sharpen focus by determining the location of edges, to track movement of objects from one visual sector to the next, to determine where objects begin and end, etc. Thus the general tendency of the cybernetic brain to order itself into a series of interrelated functional maps allows those neurons with the most to say to one another to "discuss" more easily phenomena occurring within their respective sectors. At this stage of evolution, neurons *can* accurately be described as rudimentary information processing devices. Their arrangement into maps is simply good business management strategy—just as *you* might arrange your employees so that other members of their project teams are in adjoining cubicles. The original purpose of cybernetic sensory maps is solely to facilitate this type of "intraneuronal discussion" in an efficient and orderly manner.

Sensory maps that specialize in the detection of given features are hierarchically organized, with one map projecting onto another. In a cybernetic animal this would allow for additional layers of computational analysis of primary sense data through feedback and feedforward loops. By first isolating salient features of a particular sense modality into a series of discrete maps and then allowing those maps to develop complex interconnectivity, the separate datasets can be compared and contrasted, permitting a more sophisticated analysis.

Neuroscientists call these features "maps," which is entirely appropriate when speaking of a cybernetic animal, for which these maps are simply organizational tools—completely utilitarian—like something one would fold up and tuck away in the glove compartment until needed. All the same, I find it very curious how the habitual use of a given term can severely limit the collective's thinking. By allowing the connotations of any well-established term of art unduly to influence the course of the scientific investigation of a particular phenomenon, one can easily end up overlooking a more obvious alternative view. These brain structures have always been referred to as *maps*, possibly because they were initially "mapped" out point by point, carefully and methodically, using the fine needle of a brain probe. If they had first been discovered by an entirely different method, through some as yet undeveloped technology that allowed scientists to *visualize* in infinitely fine detail the overall electromagnetic activity of these enormously complex neuronal structures *while in the act of perception*, it may have been much more apparent that these cortical areas are actually "moving pictures"—multiple venues for a succession of sound and light shows, all displayed and performed simultaneously, as if on various screens and stages throughout the "Cerebral Multiplex."

In a primitive, preconscious visual system, a bright object passing horizontally across a darkened visual field would cause a corresponndding group of photon detecting cells on the retina to fire in a sequence precisely matching the movement of the object. From there the moving image would be directly translated onto a topographic map within the primitive visual cortex, where these disturbances would undergo some form of computational analysis. An initial flurry of neuronal activity in the appropriate sector of the visual cortex would accompany every movement of the object, causing this region of the brain to appear to function in a manner reminiscent of Aristotle's *camera obscura* (a completely darkened room with a tiny aperture in one wall that causes an inverted image of stationary and moving objects in the outer world to be projected upon the opposite wall).

Just as in the ancient *camera obscura*, the image projected upon the surface of the primary visual cortex is likewise inverted, appearing upside down (to an *external* viewer). The inner experience is, of course, of a right-side up world. An inconvenient fact for our purposes—this most likely signifies another inalterable inheritance from our preconscious cybernetic progenitors. If no one is watching the TV, who cares if the set is upside down? However, experiments beginning with those of George Stratton in the 1890s show that volunteer subjects wearing mirrored goggles designed to present the world upside down will after several days begin to see the world right-side up again. Within a week, the transition is complete. After the goggles are removed, the subjects' vision is once again reversed for several hours.

Another line of reasoning may help us to see how the topsy-turvy presentation possesses its own convoluted logic. Since animals walk upright, for the most part, objects that appear lower down in the visual field—the ground under their feet, for instance—are consistently *closer* to the animal's body than those that appear in the upper portion of the visual field, nearer the distant horizon. The bottom edge of the visual field (which is presented at the top edge of the visual cortex) is therefore rightly placed *closest* to the sensory homunculus.

We are engaging in a great deal of speculation about hypothetical sensoria when we each have a perfectly serviceable sensorium right under our own nose, or rather, right inside our own cranium, to poke around in. Let's relax for a moment and then continue with our thought experiment, incorporating some valuable and incontrovertible first-person data. Settle into your easy chair, and get comfortable.

After reading this paragraph, close your eyes for a moment, and notice that the world immediately disappears, replaced at first by its inverse, in ghostly afterimages. These then dissolve into a chaotic profusion of sparkling dots. At this moment you are directly experiencing the random firings of the neurons of your own visual cortex. This is what they look like when they have nothing to look *at*—the snow on the television screen when the cable goes out. Cup your palms over your eyes, and attempt to peer into the darkness several feet in front of you. Observe the whirling galaxy at the center of your visual field that maps your hyperactive foveae, with their crowd of cones now searching for something, anything, to focus on. Note your inability *not* to try to make *some* sense of the visual world from the persistent afterimages, as you call up memory maps of the room you just left, of the book that just disappeared from view. Notice how even your *fingers* try to help you to sustain some memory of its fading visual form. Recall how this morning you reemerged into waking consciousness from a vivid dream of another world. It was your reengagement with the six sense realms that allowed you to transition back into the sensorium. Effortlessly, as if you had done it a hundred thousand times before, you reconnected with the multitude of sensory topographic maps, entering the labyrinthine structures of your brain like a hand easing into a very familiar, soft kid glove.

    Just because it's a *common* experience—("What? *Consciousness?* Oh, that old thing!")—doesn't mean that it isn't a kind of miracle. You reenact this miracle every morning, you sustain it every moment of your waking life, and you even carry this miracle of consciousness with you at night into the archetypal realm of dreams.

    Return once more to consider the ancient totem animal you conjured up for our thought experiment. Dive into its body-mind with the full force of your imagination. Locate those shimmering maps and moving pictures of your world, scattered though they may be throughout the ganglia, nuclei, and cortices of your new home. You are living the "maps" created by your cybernetic ancestors. You dove into those maps as your first refuge, out of fear and bewilderment. They sparkled and shone with the glimmering light of presence, of location, of stability—offering you the relative safety of a well-armored worldview.

    So this is the story as I see it. The body never intended to lure awareness into becoming the illusory *self* or to convince it to take up residence within sensorium consciousness. Neurons originally fired their electromagnetic charges simply to speed messages from one part of the body to another. The "light" produced by those firing neurons was at first a meaningless epiphenomenon. Brains evolved topographic maps simply as the most efficient means to organize and robotically analyze sense data for the purpose of producing more complex, adaptive response behaviors. Brains never intended to construct these earliest black-and-white precursors of the brilliant Technicolor displays that we now inhabit and enjoy. Like magic mirrors hung on the walls of an empty room with no one to reflect upon them, these maps were likewise, originally, simply biomechanical epiphenomena.

At first, in momentary flashes, the *bare awareness* inherent in the quantum field simply *became* the luminous, ever-shifting representation of the creature's surroundings. There was, initially, perhaps no sense of *self* per se, only the consciousness of *being* the world that was seen and heard—not unlike a newborn human infant who has not yet learned to distinguish itself from its mother and who views the entire continuum of sensation as one undifferentiated whole. You may contend that the first sentient being endowed with sensorium consciousness could never have been fooled by such a primitive display. Yet by way of analogy, consider how mesmerizing it must have been to hear for the first time a symphony orchestra mysteriously emanating from a scratchy wax cylinder recording on the earliest hand-cranked Edison phonograph in the late 1800s, or how the very first motion picture audiences leapt screaming from their seats at the apparent approach of an oncoming train. Think how gullible we all are to the absurd discontinuities in the landscapes and storylines of our own dreams. Awareness emerges in a blinding panic and gladly embodies whatever display will provide it with some sense of solidity and permanence. In those first moments of creation it has absolutely no sophistication—it is happy to have found even the humblest, most primitive sanctuary.

How is it that the evolution of something as miraculous as sensorium consciousness is made to appear so simple and inevitable? Can it be that upon nearly all stable "Class M" planets in the universe the light of their suns provides the energy that routinely initiates the chemical reactions that create the self-replicating molecules that make the proteins that form the cell bodies that become the organisms that develop motility and evolve into the animals that diverge into the predators and prey that require the quick-wittedness that leads to the diversification of brain states and the development of the sensory maps that become the well-lighted dwellings of innumerable sentient beings? To my mind, this is the most astonishing, multidimensional *Creation* story imaginable—of consciousness emerging from the inside out—*Light* arising from the ground of being to meet and mirror and merge with a world brought in through the eyes and ears of all sentient existence. Upon numberless planets throughout the universe, *Light* creates through this ceaseless activity the lineages of Angels.

PSYCHE'S PALACE

A SUMMARY OF CHAPTER ELEVEN

The gulf between presensorium and sensorium consciousness seems at first glance to be unbridgeable. How could something so extraordinary, so completely unlike anything else in the known universe, have arisen by means of evolutionary processes alone?

Whenever we speak about consciousness, we are by definition prying into the realm of the Great Mystery. But if we hold to our basic postulate—that some still enigmatic form of proto-awareness dwells within the quantum field itself—there is the prospect of a plausible explanation for the transition from an earlier, nonconscious cybernetic phase of animal evolution to the current era in which full-fledged sensorium consciousness self-evidently exists.

The sensory cortices of the cerebral cortex include the primary visual cortex and other associated visual cortices, the auditory cortex, and the somatosensory cortex. The primary visual cortex is a leading contender for the genesis point of sensorium consciousness, because it is laid out upon the surfaces of the occipital lobes of the cerebral cortex in the form of an incurvate retinotopic map that effectively reproduces the surrounding visual scene in a point-by-point representation, with the electromagnetic leakage from its firing neurons acting as pixellating elements on what will soon become a "self-conscious" viewing screen. In a preconscious, cybernetic animal, the area of the brain responsible for analyzing visual data would inevitably have evolved into just such a two-dimensional retinotopic map, although no perceiver of the "image" would yet be present.

With these purely "mechanical" cortical structures firmly in place, the proto-awareness inherent within the associated quantum electromagnetic field was able to become *the view from that position—only much later evolving into a bona fide self perceived to be "distinct" from the world surround.*

CHAPTER TWELVE

NOTHING DOING

Look, it cannot be seen—it is beyond form.
Listen, it cannot be heard—it is beyond sound.
Grasp, it cannot be held—it is intangible.
These three are indefinable;
Therefore they are joined in one.

From above it is not bright;
From below it is not dark:
An unbroken thread beyond description.
It returns to nothingness.
The form of the formless,
The image of the imageless,
It is called indefinable and beyond imagination.

Stand before it and there is no beginning.
Follow it and there is no end.

PSYCHE'S PALACE

Stay with the ancient Tao,
Move with the present.

Knowing the ancient beginning is the essence of Tao.

> Lao Tsu
> Tao Te Ching
> (translated by Gia-Fu Feng)

That which is body is pure body, composed entirely of matter—what today is biologically active will one day become a corpse without losing in that moment a single molecule of its form. That which is mind is pure mind, composed entirely of *Light*—what is now the brain's bioluminescence will on that day no longer be able to sustain its presence here. Body is entirely empty of mind. Mind is entirely empty of body. And yet they hold one another in the most passionate and intimate embrace, bound for this moment in a Tantric weaving—flesh and spirit commingling their separate powers to form these intricate patternings of function and display. This is the essence of nonduality. Body as body, without mind, could never have become this particular body. Mind as mind, without body, could never have become this particular mind. Body and mind have coevolved upon this planet into the absolute manifestation of *Not Two*.

> *Thirty spokes share the wheel's hub;*
> *It is the center hole that makes it useful.*
> *Shape clay into a vessel;*
> *It is the space within that makes it useful.*
>
> *Cut doors and windows for a room;*
> *It is the holes which make it useful.*
> *Therefore profit comes from what is there;*
> *Usefulness comes from what is not there.*
>
> Lao Tsu
> Tao Te Ching

NOTHING DOING

Body and mind meet upon the surfaces of one another, like the clay vessel and the space within. There is a well-known optical illusion, in which one can see either two dark faces staring at each other across an empty void or a single white vase poised on a dark background. It is impossible to see both images at the same time—the one is figure to the other's ground. So it is with the body and the mind—and this is the very conundrum that can easily keep a dedicated neurobiologist "in the dark" for many years, if he focuses all of his attention upon the *dark matter* of the brain alone. The leap of imagination that brings everything into the *Light* is made by refocusing the mind upon what is most *obvious*. However neurons may be arranged in their networks and whatever they may be saying to one another, they are making an incredible amount of *noise* while doing so! What self-respecting biocomputer would have resisted so mightily the evolutionary pressure to conserve energy by developing systems that required less energetic expense to complete each "informational" transaction?

If we compare the evolutionary development of the brain to the history of technological advancement in computer design, there are some obvious parallels and one literally glaring incongruity. In the computer industry, we can see definite signs of an "evolutionary struggle" within the marketplace to get the most bytes (or megabytes or gigabytes or *terabytes*) for the buck. From the early days of primitive vacuum tubes, through the era of transistors, into the age of microchips and central processing units, successful computer hardware designers have always followed the five sacred edicts that have kept their product on the cutting edge, insuring its "survivability"—make it more compact, make it more complex, make it faster, make it cheaper, and make it more energy efficient (nowadays not so much to avoid a high electrical bill, but to keep the CPU from melting!).

Although the first three criteria apply as well to the evolution of the brain, apparently the fourth and fifth do not. Expending precious calories by the wagonload, setting off a profligate display of fireworks within the skull, the brain looks much like an oak tree that chose to set itself ablaze. Although the brain accounts for only two percent of the body's weight, it can use up to thirty percent of the calories consumed each day. You would think that over the course of hundreds of millions of years of evolution, neurons would have succeeded in evolving methods of communication that required less electrical energy, but this does not appear to be the case. Quite the contrary. Since natural selection favors the development of efficiencies of all kinds in every other biological system, how can it be that the brain is apparently breaking this fundamental law?

Such "wastefulness" would only make sense if the primary driving force behind the evolutionary development of the brain was not the sober, penny-wise necessity that prompts the reluctant acquisition of upgraded computer hardware, but rather the delirious enthusiasm that leads one to make the pound-foolish purchase of a new home entertainment center and big-screen plasma TV. What we are shopping for in that case is a *Brighter! Sharper! More Lifelike!* display. To get those effects you need energy, and lots of it. The greater the electromagnetic field strength—the more photons—the better.

The sensorium (personified) is like the artistic director of a Broadway musical, who has to fight it out with the producers (the rest of the body) to get as big a budget for his stage production as the theater company will allow. As the producers will remind him, those calories don't come cheap. They represent a huge investment of time and energy, forcing the company to go regularly on tour in a voracious pursuit of sustenance. The artistic director's retort: give me enough funding for some decent lighting, an elaborate set, some shiny costumes, and a full orchestra, and I'll make the audience hungry for more!

If the ultimate object of the brain is to produce a bright, compelling, irresistible display, then the firing of sensorium neurons should be much brighter than would be absolutely necessary simply to send "information" on to the next cells down the line. In fact, a statistical analysis of neuronal efficiency would be the surest means of establishing which neurons are involved in sensorium presentation. Those whose prodigal energy usage cannot be justified in terms of information processing efficiency must somehow account for their having escaped the strong evolutionary pressures to economize.

One particular region of the neuronal cell body would act as the strongest and the most focused "pixellating" element for sensorium presentation. Between the cell body and its axons is the *axon hillock*, which boasts an extremely high density of sodium channels in its cell membrane. By means of the rapid opening and subsequent closing of these voltage-dependent channels, positively charged sodium ions are instantaneously pumped out of the cell body of the firing neuron—thus generating a spike of depolorizing current. The axon hillock is therefore the site of the largest concentration of electric charge during neuronal firings. Accordingly, it is also the point at which the greatest fluctuation in the electromagnetic field occurs. It is important to bear in mind that the seemingly "excessive" electromagnetic activity at the axon hillock cannot cause the signal (the traveling action potential) to move any faster down the axon, nor does it in any way affect the type or quantity of neurotransmitter released by the axon terminus at the synapse. No information is being transmitted between neurons by all of that electromagnetic "shouting" at the axon hillock. The signal is actually passed on by a quiet handoff of neurotransmitters spilled out from tiny vesicles at the synaptic junction.

This new paradigm has required us to look at light and electromagnetism in a new way. I have given its conscious, spiritual aspect a typographical upgrade by consistently referring to the phenomenon as *Light* (italicized and capitalized). Humdrum light (unitalicized and lower case) refers to its pedestrian role as flashlight beam or flickering candle flame, the plaything of optical physicists. *Light* merits the exalted status of being the essence of consciousness only after having organized and transformed itself within the incredibly complex and refined biophysical structures of the brain.

To grasp this fundamental realization—that worldly light reveals its spiritual nature directly within your own mind's eye—is to understand the essence of nonduality. Here spirit and matter are so thoroughly enmeshed that they function as an apparent unity. Descartes's *res cogitans* and *res extensa* (the "thinking thing" and the "extended thing"—which reified the foundational split in Western civilization between spirituality and philosophy on the one hand and science on the other) are still as distinct as they ever were, but his conclusion that this implies an impassable gulf between them was never truly justified. I refer to this paradoxical view of the intrinsic distinction and equally intrinsic inseparability of spirit and matter—mind and brain—with a nod and a wink as Neo-Cartesian Nondualism.

The sacred teachings from a wide variety of sources converge on this particular point. Advaita Vedanta (the nonduality schools of Hinduism), both Mahayana and Theravadin Buddhism, Chinese Chan, Japanese Zen, Tibetan Dzogchen, Taoism, Sufism, Jewish Kabbalah, and mystical Christianity—all state unequivocally that the ultimate nature of reality is *nondual*.

Nondual means simply "not two." Whereas duality may be expressed as being and nonbeing, *self* and other, *samsara* and *nirvana*, good and evil, illusion and reality, action and doer, spirit and flesh, *Brahman* and *atman*, God and man—nonduality points to a realization beyond all logical categories, but not opposed to them—one that joins all philosophical opposites into a functional whole without obliterating the uniqueness of either side. Even nonduality itself is nondual! Mahayana Buddhism adds this twist by positing the nonduality of wisdom and compassion—where wisdom is defined as the clear and penetrating insight into nonduality (the "emptiness" of the *self* and all composite things)—and compassion is defined as the heartfelt understanding of the *absolute* reality of the inner *experience* of duality (for those sentient beings still held captive by an unshakable belief in "*self* and other.")

Dzogchen avoids the infinite regress by referring to nonduality simply as the natural state of the mind. Advaita (from the Sanskrit *a-*, "not" and *dvaita*, "two"), whose best known modern proponent was Ramana Maharshi, advocates realization of the Self, which perceives no individual ego and does not regard itself as the doer of actions. Taoism calls this *wu wei* (from the Chinese *wu*, "not" and *wei*, "to do"), translated variously as "nonaction," "without ado," or "nothing doing"—action performed without the I-thought.

I certainly cannot hope to do much more than scratch the surface here. Volumes of ecstatic poetry and scholarly prose on the subject of the experience of the nondual state are widely available and well worth pursuing. I believe that the ubiquity of this insight in so many religious, philosophical, and spiritual traditions reflects a deep intuition into the actual mechanisms of the mind.

The physical body has produced a brain comprised of a hundred billion very real, very tangible, fleshly neurons, engaged in complex real-time calculations, forming, at one level, what can only be described as a vast biological supercomputer—but a subset of these neurons are also functioning as the structural members that support something truly "not of this world." *They* determine the anatomy of the lightbody and give it its functional form. Held in place by these luminous cytoarchitectural landmarks, the lightbody embraces the physical brain, achieving something very close to unity with it.

The final display—of sight, sound, smell, taste, touch, memory, dream, language, and concept—looks, sounds, and feels as it does because somewhere in the brain a set of neurons has mimicked that particular effect in its firing patterns. The shapes and textures of these "objects" are all highly evolved stage illusions, intricate special effects perfected over millions of years of evolution. When we ask ourselves, "But how is it that these effects seem so real? How can my brain produce something that mimics reality so perfectly?"—we don't realize the tautology inherent in our question. We live and breathe entirely within a world of illusion. That is all we ever know. The illusion *looks* exactly like reality, for the simple reason that the illusion *is* our only reality. There is nothing else with which to compare it.

What causes us to neglect this self-evident truth is the fact that all six sense realms interpenetrate and cross-reference one another. To confirm the reality of a visual object, for instance, we can tap it and hear the sound generated by our touch and compare those sensations to similar objects existing within our memory. A healthy brain in a state of relaxed waking consciousness has no ulterior motive; it seeks to present the most accurate, lifelike presentation possible of the sensorium surround. Only within the grip of the most extreme emotions or when we purposefully toy with our own perceptions—by introducing psychotropic drugs into our system or otherwise inducing an altered state of consciousness—do we realize how loose our hold on reality actually is. When a drastic change occurs in the brain, whether through temporary intoxication, a degenerative brain disorder, or severe head trauma, our experience of *self* and world shifts dramatically. The brain is where *we* live. The world is where the *body* lives. We simply cannot have a direct experience of the world unmediated by the brain (at least not under normal circumstances, and paranormal circumstances are a bit beyond the scope of this book). This is not bad news, by any means. When we drop the pretense of Being Somebody we may realize that we are in essence the Nobody, the Nothing Doing, the *Light* of the primordial nonduality—the One Intimate Embrace.

Innumerable reports of near-death experiences from diverse cultures describe a brilliant, clear radiance appearing to the individual—before he is revived from the state of clinical death—and voices urging him to "go toward the light!" The return to the mystery of *Light* at the moment of death is the matching bookend that mirrors the birth of consciousness—as a flash of the *Light* of awareness arising from the plenum void. Let's read from an ancient Tibetan text—a handbook for attaining liberation at the time of death, or failing that, for successfully navigating the intermediate stages between rebirths. Sections of this text are read to a person lying in state during the fortnight after his death. These are the "Instructions on the Symptoms of Death, or the First Stage of the Chikhai Bardo: the Primary Clear Light Seen at the Moment of Death" from The Tibetan Book of the Dead:

> *O nobly born, listen. Now thou art experiencing the Radiance of the Clear Light of Pure Reality. Recognize it. O nobly born, thy present intellect, in real nature void, not formed into anything as regards characteristics or color, naturally void, is the very Reality, the All-Good.*
>
> *Thine own intellect, which is now voidness, yet not to be regarded as of the voidness of nothingness, but as being the intellect itself, unobstructed, shining, thrilling, and blissful, is the very consciousness, the All-good Buddha.*
>
> *Thine own consciousness, not formed into anything, in reality void, and the intellect, shining and blissful,—these two,—are inseparable. The union of them is the Dharma-Kaya state of Perfect Enlightenment.*
>
> *Thine own consciousness, shining, void, and inseparable from the Great Body of Radiance, hath no birth, nor death, and is the Immutable Light—Buddha Amitabha.*

In the following two chapters I describe what the U.S. Patent Office would refer to as the *preferred embodiment* of my little invention *(An Improved Mechanism for the Melding of Mind and Brain)*. It is an attempt to account for the subjective experience of sensorium consciousness through an analysis of the large- and small-scale architecture of the brain. It will likely prove to be incorrect in many of its details, but I have faith in the feasibility of the overall design. Had I been asked to devise the human brain from scratch—knowing what I know, supposing what I suppose to be true, and ignorant of all the rest—this is what I would have submitted to the creation committee.

NOTHING DOING

A SUMMARY OF CHAPTER TWELVE

Taking a closer look at the dynamic relationship between the gross matter that constitutes the physical body and the energy field of the conscious mind, we find that they are bound in a most intimate interweaving that hints at an underlying nonduality.

The seemingly lavish expenditures of caloric energy within the brain are seen as strong evidence that consciousness requires a bright, energetic display. The sentient brain's chief function is presentation, and its efforts must result in a completely persuasive construction of self and object reality.

The illusory visual and auditory presentations look and sound entirely "real" to us primarily because we have no other standard of reality with which to compare them. The conviction that there must exist a real and genuine, reliable and consistent "reality" behind the perceptions derives from the coherence of these six interpenetrating, three-dimensional sense realms. They all correlate with one another in a complex multisensory web, reifying and substantiating whatever objects and environments we encounter.

We dwell entirely in the energetic field of presentation. We are, in this sense, beings of Light immersed in a world made up of our own dazzling substance.

CHAPTER THIRTEEN

EMBRACING THE GENTLE MOTHER

In the fifth chapter of Jason W. Brown's seminal book, *The Self-Embodying Mind: Process, Brain Dynamics and the Conscious Present*, he gives an engrossing account of how consciousness arises in the brain. Chapter sixteen of *this* book will offer an in-depth look at that material, which, as you will see, is both meticulously well-reasoned and utterly revolutionary. His arguments are so compelling that one is left with the impression, upon completing that fifth chapter, that all conceivable questions have been thoroughly addressed and that the nature of the *self* and its relation to the world has been completely elucidated.

Yet there is not even a hint of speculation there—nor anywhere else in his book for that matter—as to precisely what this consciousness, whose functioning he describes so fully and so well, might be *composed of* or how it is rendered into a form discernible to itself. Not that Brown is alone in building such elaborate structures suspended in midair. This is the universal omission that consigns such theories to the field of *philosophy of mind*, regardless of how solid the underlying science is. What is it about his writing, though, that makes for such a persuasive and satisfying read?

Well written expository prose such as his should always be carefully scrutinized. And when we do look closely at his arguments, we notice that there is, perhaps unwittingly, a kind of shell game going on. A lot of deft movements and sleight-of-hand, but in the end—where is the pea? This is a two hundred page book—presenting one of the most comprehensive, neurobiologically sophisticated arguments I have ever read, based upon careful analyses of the functional properties of real brain structures, in complete accord with evolutionary theory, and verified by decades of clinical observations—and yet, not even the vaguest suggestion is given as to just what this otherwise thoroughly explicated consciousness might be *made of*. Presumably, it has something to do with the activity of neurons. So does this mean that consciousness is *information* or *pattern* or some form of *emergent complexity*? What is most intriguing to me is that such a detailed and seemingly complete description of consciousness can be made without once referencing this "missing fundamental."

Then I realized how the trick is done—that the luminous quality of consciousness has entered into these passages in the same way that spirit is infused into Shakespeare's sonnets—through the magic of rhythm and sonority, skillful imagery and evocative vocabulary. Such artistry is capable of engaging this imaginal property of mind so thoroughly that *it* provides the very element glossed over in the presentation. We don't notice that anything is missing from the theoretical model because as we read we ourselves are contributing the illumination that completes and validates the process being described. These two elements—the words that structure the movement of thought and the luminous mind that sews them together—combine to make a seemingly seamless and thoroughly satisfying whole.

It takes one, as the saying goes, to know one. I certainly recognize how much *I* depend upon evoking the fundamental mystery in you, gentle reader, when I ask you to consider the theoretical models I am presenting in this book. Strangely, with no other parallel in all of the universe, we must evaluate any conjectural theory of consciousness from *within* the very object itself. So there will come a point in time where the words must dissolve into the substance that sustains all words, images, ideas, perceptions, and finally, our "hold" on reality itself. And you are being asked to notice what it seems those words and images are dissolving into. I call it *Light*—wherein the figurative and the literal senses of this word finally exhibit their intrinsic non-duality.

It's all fancy talk—I know that. You know it, too—it's no secret. Nothing I nor anyone else can say, regardless of how many credentials we do or do not have tacked upon our walls, no matter how many reams of data tables, charts, and diagrams purport to support our positions, nothing can nor should convince you to neglect your own experience. You are the expert witness of your own mind. What you believe to be true about your own mind makes all the difference in the world in terms of what your experience of the world will be. What you believe to be true about your own mind, however, will not make the slightest difference in terms of what your mind actually is. If what it actually is and what we believe it to be could come into perfect alignment—Ah! Now that's the Holy Grail in the quest for consciousness.

No different than anyone else who has attempted a piece of expository writing, I must admit that I am trying to get you to come around to my way of thinking—because I have found there a great deal of joy and a new appreciation of the transparency of thoughts, perceptions, and feelings as I live from within what has become a very light and unburdened perspective. When you consider the extensive reach of the theoretical construct itself, it may sound heavy and complicated. But it is only complicated in the telling of it, not in the living of it.

Okay. Enough of this shell game. Push has come to shove. Just where, Mr. Holmes, do you suppose this presentation of *Light* to be occurring in the brain? Usually, those who engage in this sort of speculation visualize consciousness as something very, very tiny, compact and powerful, and beaming from some specific location—often from within the pituitary gland (behind the "third eye") or tucked away in some other deep fissure like a crane operator in his cabin. For me, the most elemental quality of *Light* is its expansiveness. It wants to enter the world, not hide away in the miserable safety of a bunker. I believe, therefore, that the body would have evolved neuronal mechanisms that allow consciousness to expand to fill the entire body, forming a vast bioenergetic field, whose radiance suffuses even the space beyond the form. Those who have experienced the subtle *chi* energies that are brought into conscious awareness through the practices of *Chi Kung (Qigong)* or *T'ai Chi Ch'uan (Taijiquan)* will attest that these fields appear to be circulating throughout the whole body, along the very meridians used in Chinese medicine. The efficacy of the Chinese healing arts is indisputable; yet their punctilious anatomical model—the plastic "human-pincushion" in the acupuncturist's waiting-room, with its precision map of labeled points and color-coded meridians—does not conform well to the gross anatomy of the body proper. This we take as evidence of a subtle, metaphysical dimension behind this healing art. It may, however, turn out to be a perfectly precise *physical* model of the micoanatomy of the somatosensory homunculus, where the healing actually takes place! Likewise, I casually surmise, do the chakras have radiant neuronal correlates ridging the apex of the brain.

Certainly *within* the brain, consciousness has always sought to expand its territory—first learning how to read from afar, then becoming familiar with, and finally diving headlong into the many glistening sensory "maps" at its disposal. Therefore, I believe that consciousness would quite simply be unable to resist making its primary home the rolling hills of the entire cerebral cortex, emblazoned as it is with its carpet of sparkling wildflowers. Yes, the "pea" of consciousness is right there under the walnut shell.

Indeed, the brain does look very much like a walnut when you carefully crack open its shell, even in the detail of its being covered with a delicate husk—which in the brain takes the form of an incredibly thin, transparent sheath only three cells thick called the *pia mater*, the gentle (or faithful) mother. The cerebral cortex (from L. "bark of a tree") is the gray matter that forms the "rind" of the cerebral hemispheres, which are covered with convolutions that faithfully cling to the pia mater. The convolutions are a most remarkable feature, one that demands a rational explanation. The brain is obviously a three-dimensional object, but in the cerebral cortex it appears to have done everything it can do to maximize its two-dimensional surface area.

What lies beneath the cortex is the brain's white matter. It makes up the bulk of the cerebrum and is composed mainly of myelinated nerve fibers, the vast majority of which (98.6%) connect one cortical region to another or to extra-cortical structures within the same hemisphere. The rest (1.4% or about 100 million axons) connect the two hemispheres through the corpus callosum. The whiteness of white matter is due to the color of the myelin that insulates the nerve fibers.

What lies above the cerebral cortex is also quite intriguing—and perhaps not an insignificant component of the presentation apparatus for the production of sensorium consciousness. The entire cerebral cortex is bathed in a clear liquid called *cerebrospinal fluid*. This fluid is contained within structures called the *meninges*, which will be described in more detail at the end of the next chapter. The pia mater is the innermost layer of the meninges and forms a transparent membrane that stretches around the entire outer surface of the cerebral cortex, and everywhere—even within the deepest folds of the cortex—the pia mater is in contact with this transparent liquid medium.

The human cerebral cortex has a highly convoluted surface. If you can picture the *gyrii* and *sulci* (the hills and valleys) unfurled and the surfaces ironed out into two sheets—one for each hemisphere—what you would see would be two vast expanses (six and a half square feet in all) of contiguous surface area upon which would be visible the tops of a multitude of tiny cortical columns, packed like a new set of unsharpened colored pencils in an immense pencil box, that encode (and, I contend, display) the full range of features that comprise the various sense modalities of vision, hearing, touch, proprioception, etc.

Ironically, in a seminal act that effectively blinded the collective imagination for centuries to come, the naked eye of an early anatomist first looked upon the substance of the cerebral cortex and, distinctly unimpressed, named it "gray matter." If he could have zoomed in upon a living brain, with subtler vision capable of seeing the full electromagnetic spectrum, he would have dubbed it more accurately and with far greater reverence the *rainbow cortex*, for that's exactly what it is—a spectacular, luminous display. The cerebral cortex is hardly *gray!* It is really much more like a delicate, shimmering, brightly colored, finely crafted silk Persian carpet (that has been wadded up and stuffed into a football helmet). The cortical columns would be analogous to the individual strands of monochromatic silk yarn that "pixellate" to form the intricate geometrical patterns that appear upon the carpet's surface.

The cerebral cortex is a device for the production of electromagnetic *qualia*. A *quale* (the singular, pronounced KWAHL-ay) can only be defined tautologically or by analogy: "The 'redness' of a flower is what gives it its... well, redness," or "The flower's 'redness' is like heat, like the warmth of a fire." Qualia are the innate properties of experiences that determine exactly what it is like to have those experiences—

the purpleness of purple, the sweetness of sugar, the brassiness of a bugle call. Qualia are ineffable—they are the internal bridge between the experience and the experiencer. Qualia are the deeply embedded attributive adjectives from which we infer an underlying noun. They put the real in reality.

Whether qualia are in essence real or illusory phenomena is a question hotly debated in philosophical circles—philosophical circles being self-defined as whirlpools of endless debate. Real *and* illusory is what they are, of course. The soul, in its electromagnetic manifestation, is self-evidently capable of perceiving certain vibrational realms of experience, which are segregated during normal business hours into the discrete sense modalities of vision, hearing, smell, taste, touch, and thought. Each modality has a separate palette of qualia appropriate to the vibrational characteristics of that modality. Throw in a few recreational chemicals after hours (or speak to someone with the neurological condition known as *synesthesia* who routinely "hears" colors or "sees" sounds) and you will understand how the presumably solid walls between these sense modalities are in actuality only thin veils of convention. The soul's gift is its ability to hallucinate the forms of its own creation, which it fully embodies as a field of dreams. The body's gift in turn is its ability to present a rational program to this giddy scintillation and convince it that it is what it tells it it is. The task of interweaving these separate "its" into a convincing sensorium experience, I contend, falls to a class of *presentation neurons*, most notably those of the cerebral cortex.

The microanatomy of the cerebral cortex reveals structures that seem ideally arranged for presentation. Almost the entire cerebral cortex is covered by the same sequence of six distinct layers of neurons and associated dendrites and axons arranged in planes locally parallel to the pia mater, in total two to five millimeters thick, often subdivided into additional layers—varying in their precise organizational details according to the special needs of each cortical area. In addition to this *laminar* (or *layered*) organization, the cerebral cortex exhibits *columnar* organization as well. Throughout the cortex a similar structure is found: orderly arrays of *cortical columns* (just like the silken threads of the carpet), made up of approximately 100 neurons, passing straight through these layers of cortex, all perpendicular to the plane of the pia mater, and engaged with it at a single point, a *pixel* if you will, at the outer surface. Within cortical columns, all neurons are functionally linked and coded for a particular sensory feature, such as, in the visual system: direction of motion, angle of orientation of edges, color, and depth of field.

Separate auxiliary visual areas (known collectively as the extrastriate visual cortex)—which include the secondary, tertiary, quaternary, etc. visual cortices, also called V2, V3, V4, etc. up to V8—receive a point-by-point mapping from the primary visual cortex (V1). Each area appears to be responsible for processing one or more of the submodalities of vision. In both V1 and V2, there is an organized array of color sensitive subcolumns known as *blobs* that receive their input from the three types of retinal cones, whose maximum sensitivities are to wavelengths of 440, 520 and 580 nanometers—corresponding to blue, green, and red light. These blobs are surrounded by *orientation columns* (each sensitive to a particular angle of orientation), which in turn are nested within even larger *ocular dominance columns*, whose functioning will be described in more detail in the following paragraphs. Automatic processing between areas V1 and V4 permit the property of "color constancy" to emerge, since the neurons in V1 are sensitive to wavelength, whereas those of V4 are sensitive to absolute color. This comparison allows colors to be interpreted in the context of surrounding colors, thereby "discounting the illuminant" so that changes in lighting conditions do not radically alter the perception of an object's color.

When analyzing the structure of a given cortical map or *presentation screen*, one needs to bear in mind that there are always two distinct conceptual manifestations of neuronal "connectivity" to be considered: one for the brain and one for the lightbody. The first manifestation is the only one that has been deemed worthy of study by neurobiologists to date, since it is—from their perspective—entirely self-evident that only within the *axon-dendrite connectivity* can the all-important computational information processing occur.

The unexamined manifestation is actually the more obvious one, the *image connectivity* based purely upon neuronal proximity and geometrical brain-space relationships. The sensorium can only be created by virtue of the actual physical arrangement of neurons upon these planes of presentation and the appearance of their resultant collective firing patterns. The crucial point to be borne in mind is that this image connectivity is entirely distinct from the computational axon-dendrite connectivity that may (or may not) interlink neurons adjacent in brain-space.

The two forms of connectivity are always superimposed upon one another, but each performs a separate function, and each reports to a different master. Image connectivity considers each neuron solely in terms of its physical location on the grid. Neighboring neurons are "connected" in this way, by the simple fact of their being next to one another in brain-space, regardless of whether or how richly they are interconnected dendritically.   This is the source of the apparent "image continuity" of the electromagnetic presentations (whether of visual imagery, tactile sensations, auditory pitch relationships, etc.) that flash across the surfaces formed by the vast matrix of neuronal cell bodies. An understanding of the theory of neurobioluminescence in the evolution of sensorium consciousness hinges entirely upon this one conceptualization: that it is these neuronal structures alone that form the completely material bodily framework that serves to stabilize the completely *immaterial* lightbody.

In this model, neurons do not send packets of "consciousness" to one another along their axons. What they *are* sending may be bits of information that allow for the correct presentation of the "lighting effects" that are subsequently perceived by awareness, or they may in other instances be operating beneath the surface of presentation as elements of a passive medium, arrangements of circles within circles rhythmically resonating complex patterns of light or sound.

Since the axonal-dendritic connections form structures that are in this important sense completely independent of presentation structures, the hidden layers of purely computational brain activity have free reign to manipulate the display. The *perception* of the meaningful patterns of *Light* produced by these cortical maps is entirely dependent upon the spatial orientation, proximity, spectral "color" and luminosity of their constituent neuronal cell bodies, whereas the *execution* of any particular lighting effect depends upon computational factors—the specific layout of the neurons' "electrical wiring" diagram. Not all of the neurons need be involved in the *display* of the lighting effect. A given cortical area may enlist many more neurons beneath the surface to "strategize" the effect than are actually involved in producing the patterns of *Light* by their sequence of firings.

This helps elucidate an important distinction between the display mechanism and the displayed content. The difference between the lightbody and the brain is somewhat akin to the difference between the illumination upon the surface of a television screen (the image that actually grabs our attention) and the television set itself, the DVD and DVD player all hidden within the cabinet. The image and the mechanical devices work in tandem—one would not be there were it not for the other—but they are essentially organized quite independently of one another.

I refer to this independence/interdependence of brain and lightbody, matter and energy, form and function, reality and illusion rather light-heartedly as Neo-Cartesian Nondualism. The split between spirit and matter is still absolute and irreducible, precisely as Descartes pronounced it. However, the unity of spirit and matter is absolute and irreducible as well. Both have evolved in synchronous orbit with one another, ever since the first prehistoric animal was initially "animated"—endowed with *anima*, or soul.

The theory of sensorium consciousness attempts to tease apart a great number of overlapping biophysical, bioenergetic, mental, and spiritual processes and structures; it is not a simple task to wrap one's mind around the concept in its entirety. If the neuronal activity of the primary visual cortex, for example, is analogous to a brightly lit television screen that we simultaneously *are* and are *watching*, then this is one extraordinary machine—highly technologically advanced—unlike anything on the market today. Behind the screen of this TV set is an extremely sophisticated computer that minutely analyzes all aspects of the scene you are being presented with—heightening the realism of the images, analyzing the ambient lighting to adjust for color constancy, drawing stark edges around objects, pulling them from the background into three-dimensional space, determining their identities, researching their past personal associations, and infusing them with an appropriate emotional affect. Relevant tie-ins to other sense modalities are instantaneously displayed on an array of adjunct presentation structures. Welcome to the Neo-Cartesian Theater-in-the-round—where featherlight, state-of-the-art virtual reality headgear inwardly broadcasts brilliant flashes of light and furious sound to captivate utterly its unwitting audience of *one*.

A SUMMARY OF CHAPTER THIRTEEN

The two hemispheres of the human cerebral cortex have surfaces that are highly convoluted with gyrii and sulci (hills and valleys), but essentially the hemispheres can be conceptualized as a pair of contiguous two-dimensional sheets upon the upper surfaces of which are exposed the tops of a multitude of tiny cortical columns that encode and display the various "qualia" appropriate to the particular sense modality—be it vision, hearing, touch, taste, smell or proprioception.

The neurons within the sensory cortices of the cerebral cortex are simultaneously performing two distinct functions. The first is physical and computational, involving the complex, long- and short-range axonal and dendritic connections that allow sense data to be transmitted, analyzed, and prepared for presentation. The second is purely presentational, whereby the spatial relationships between adjacent neurons, their firing patterns, luminosities, spectral color, and other bioluminescent effects are showcased.

If there is a single concept that distinguishes the theory of neurobioluminescence in the evolution of sensorium consciousness from all other contending theories of consciousness, it is this: there are always two forms of neuronal connectivity that need to be considered when analyzing the function of sensorium presentation in a particular sensory cortical area. Neurons are richly interconnected by their axons and dendrites, and these connections provide the means for the brain to accomplish all of its complex computational and organizational functions. In contrast, the presentational "image connectivity," considers each neuron only in terms of its location on the cortical surface grid. Neighboring neurons are "connected" by the simple fact of their being next to one another in brain space. The first form of connectivity is relevant only in terms of the functionality of the physical brain. The second form of connectivity is meaningful only to the indwelling consciousness.

A detailed discussion of relevant cytoarchitectural features of the brain and some theoretical conjectures concerning their possible functions are broached here and in the following chapter.

CHAPTER FOURTEEN

FURTHER CONJECTURE

To demonstrate how the present theory is able to reduce a complex neurobiological conundrum to what may well be its self-evident solution, consider the following example from the primary visual cortex. How binocular vision produces the illusion of three-dimensional space is a classic example of the "binding problem." How do these two discrete sources of information combine to form the illusion of dimensionality? If we theorize based solely upon surface *appearances*—of what each hemisphere's half of the primary visual cortex would be *presenting* to consciousness through the firing of its neurons—we will see quite clearly how the basic three-dimensional display is both created and *experienced* independently in each hemisphere. With the lightbody directly engaging the surfaces of the primary visual cortex, no further input or calculations would be necessary for it to immerse itself in a presentation of three-dimensional reality—one entirely consonant with our own day-to-day experience.

In both hemispheres, the primary visual cortex is mottled by tiny *ocular dominance columns*—each of them one half to one millimeter across—that provide each hemisphere with an equal sampling of visual information from both eyes. These two "dappled" images are superimposed upon one another, with each eye providing half of the total illumination. Although the two images are broadly the same and readily combine, the slight difference in perspective between the left eye and the right eye is what allows binocular vision to create the first approximation of three-dimensional space in the sensorium. The twelve extraocular muscles allow the eyes to adjust their separate *lines of sight* to intersect at a single point where the object of focus is located in a three-dimensional scene, thus determining the momentary *plane of focus*, a limited area extending laterally from the *point of focus* and perpendicular to those approximately parallel lines of sight. The lines of sight are conceptualized as two imaginary rays originating at the fovea centralis (the point of focus at the back of each eye) that pass straight through the center of each pupil and continue directly outward to intersect at the point of focus. These lines of sight are parallel when the object of focus is at infinity.

If a second object is beyond or in front of the plane of focus, its image would be literally doubled upon that hemisphere's primary visual cortex (appearing upon the visual cortex itself and therefore *experienced* as two less solid, transparent, "ghostly" images, each partially merged with the background), whereas the two images of an object *within* the plane of focus would naturally meld into a single, solid image, experienced as such upon the visual cortex. When the two views are interlaced in this way upon each separate hemispheric screen (due to the randomly placed ocular dominance columns), the conscious mind is able to determine the plane of focus by simply noting which of the many objects is solid and singular, not ghostly and doubled. The farther away from the plane of focus a second object is, the farther apart its doubled images would be from one another.

Because of the direct mathematical correlation between the spacing of the doubled images upon the visual cortex (measured in millimeters or fractions of millimeters) and the actual relative distance of the object itself from the eyes and from the exterior plane of focus (measured in inches or yards), these doubled images are perhaps the most important clues the brain *and* the lightbody have in determining the precise position of an object in three-dimensional space. The brain does not calculate locations in space based upon the angle of inclination of the eyes—unlike a computer, the brain has no need to use complex *vector analysis* to navigate within the third dimension.

FURTHER CONJECTURE

The geometry of three-space is translated *directly* onto each of the the two-dimensional cortical surfaces, the *third* dimension being precisely indicated by the distances between doubled images. If there is no distance between them, if they appear as a unified image, then that defines the zero point, the momentary plane of focus. The greater the separation between the doubled images, the farther the object must be from the plane of focus.

Determining whether this "doubled object" is behind or in front of the plane of focus requires some additional contextual information. Further neurocomputational processing to compare the measured distances between various sets of doubled images upon the plane surface of the visual cortex would allow for a more accurate three-dimensional map to be configured and displayed within the sensorium, perhaps directly by the cortical columns that code for relative depth of field. Just how that subtle quale of depth (the feel of distance) is made manifest to the lightbody by these particular cortical columns, I cannot yet say—but by the end of this chapter I may be willing to hazard a guess.

Surprisingly, most of us rarely notice that every object in binocular visual space that is not located within the thin plane of focus is doubled in this way. People are often quite surprised when this is pointed out to them. Let me demonstrate. Point the index fingers of both hands at one another so that the tips are approximately a half inch apart. Hold your hands up in front of your face in this position and focus on your fingers. Nothing special. Now raise your eyes just over the top edge of your index fingers and focus on an object in the distance. You should see a flesh-colored "sausage" floating mysteriously between your fingertips. At any given time, the vast majority of objects that appear in the visual field are similarly doubled. By constantly shifting the plane of focus, scanning various objects in the foreground, middle ground, and background, we prove to ourselves that all of the objects are indeed single and solid. This troubling doubling is usually completely ignored as we deftly wander through three-dimensional space during the normal course of our day.

In fact, if one is willing to put in a bit of effort at the beginning and risk a possible initial headache as one accustoms oneself to the technique, one can learn how to "pop" a two-dimensional photographic or video image into "three-space" within the mind. The trick is to cross one's eyes ever so slightly and to soften one's focus, while taking in the full periphery. Now *every* part of the image is doubled—the picture's actual physical surface no longer defines the plane of focus—so the mind must rely upon other clues to model the three-dimensional image. It also helps if one is able to "suspend disbelief" by relaxing into a slightly more meditative state of mind. One must convince oneself that one is actually able to look "through" the surface of the screen or the photograph—in order to open up enough mental space for the mind to expand into. Eventually one sees everything in the image as if it were projected into three-dimensional space. The effect is astoundingly real! Very complex objects—the foliage of trees, individual strands of hair blowing in the wind, the angled planes of buildings in crowded cityscapes—are all rendered perfectly into a three-dimensional image. This is especially fun to do during action sequences in a large-screen movie theater. Without any special projectors or flimsy two-tone cardboard glasses, one is able to experience the movie as if it were shot in 3-D.

This should not be all that surprising when you consider the evolution of the visual system. Two-dimensional images are in fact the anomaly; they never occur in the natural world. It is surprising that we have no trouble recognizing them as representations of three-dimensional objects—a tribute to the power of the human imagination. Other animals, apparently, are not so easily fooled. Such realistically rendered images have only been around for at most a few thousand years. The revolutionary advancements in painting technique that occurred during the European Renaissance were essentially an exploitation of the hidden powers of the brain. These paintings are capable of inducing subtle effects in the visual cortex that stimulate *its* palette of electromagnetic "pigments" to render an illusion of dimensionality somehow more "real" than life itself. Through the use of precisely accurate or dramatically exaggerated geometrical perspective, *chiaroscuro* (the play of light and shadow), and *sfumato* (the illusion of distance created by depth of color that mimics the thickening atmosphere), an array of techniques were developed to take full advantage of the brain's innate capacity to model three-dimensional space.

To enhance the realism of the three-dimensional sensorium, the brain's subconscious special effects departments rely upon receiving hundreds of subtly distinct perceptual cues. Remember that these "pixels" in the visual cortices are not passive like the pixels on a television screen—they are richly interconnected with a multitude of active information processing submodules. The visual cortex employs every gift it has inherited from its ancestors and every skill it has learned from its years of experience to hone and polish the final presentation of the visual sensorium. Whatever qualia the soul is capable of experiencing (that can be matched with observable real world phenomena) the body has learned to employ in refining its electromagnetic representations of the three-dimensional world.

The sense of vision has quite a repertoire of talents to draw upon: its accumulated knowledge of the effects of lighting and shading and the spatial arrangement of cast shadows; its understanding of the likely size range of various objects, how objects appear smaller in the distance, the way in which distant objects may be partially occluded by objects nearby, and how objects become progressively more blurred the farther they move from the plane of focus; its capacity to compare current visual information with the memory of the visual properties of similar forms; and its ability to analyze the velocities and trajectories of moving objects (and to use a similar analysis to account for the *apparent* movement of stationary objects as the *observer* travels through the landscape).

All of these processes in concert allow the visual cortex to perform its primary function, which is to construct a full and accurate three-dimensional representation of the surrounding space that is experienced as a *real world* by the lightbody. Binocular vision is not the only mechanism responsible for the creation of three-space; relying entirely upon these other, ancillary visual cues, a one-eyed person is still able to reconstruct in his mind an only slightly diminished three-dimensional world. As a matter of fact, a much simpler way to experience the three-dimensionality of a two-dimensional image (as described above) is to cover one eye completely and allow the mind itself to drift into the soft mental focus of three-space. The effect produced by the technique of slightly crossing one's eyes—in my own experience, at least—produces visual imagery that seems much more immediate and realistic, while the technique of covering one eye produces an image that appears more distant and pictorial.

In any attempt to validate the theory of neurobioluminescence in the evolution of sensorium consciousness, one of the first steps would be to determine which of the cortical and subcortical topographic maps are actually *experienced* by consciousness and which ones are simply their associated, *nonconscious* sense information processing substations. Never shy about engaging in wild speculation on these matters, I have given some thought to a particular pair of structures—the lateral geniculate nuclei—which are two small oval masses located in the thalamus (part of the diencephalon, in the lower central brain, above the brainstem). They are situated at the termini of the optic tracts, the name for the optic nerve fibers that extend beyond the optic chiasm (the point of crossing of half of the fibers of the optic nerves from each eye). I have wondered how an aspect of the lightbody might actually operate within this intriguing structure.

When speculating about how a particular cytoarchitectural feature of the brain might contribute to the sensorium experience, a thought experiment is always useful. Simply place yourself within the structure—in your imagination—and visualize how the electromagnetic field produced by the firing of its neurons might appear from that location. Let's consider one of the two lateral geniculate nuclei—the one in the left hemisphere—as an example. Electron microscope images of the LGN, viewed in section, reveal that it is shaped like a three-dimensional amphitheater, with six concentric projections from the left half of the visual receptive field arranged like tiers of seats around a central stage. My first inclination would be to situate myself directly in the center where the "actors" would be, but reversing roles with the "audience," face outward to perceive what are in fact six concentric, overlapping *projection screens*. There I would notice that the first, fourth, and sixth layers of moving images are presenting nearly identical versions of the visual receptive field from the left half of the retina of the right eye. The second, third, and fifth layers interweave similar displays, but from the left half of the retina of the left eye. The first and second layers are made up of large-bodied cells that respond only to light and dark, which would create a black-and-white moving picture on the first two "screens." Small bodied cells in layers three to six mediate color vision, "tinting" the black-and-white movie.

From this very scant evidence alone I would conjecture that the lateral geniculate nucleus may well function as an *environment* in which an aspect of the lightbody resides and from which it is able to "position" multiple objects in the foreground, middle ground, and background of the visual field. Distinctions between what the left eye and the right eye see would allow such a *perceiver*, located directly in the center of the LGN and sensitive to all six presentation screens, to establish a first approximation of the relative locations of objects in space. The fact that the LGN is so radically incurved (much like the primary visual cortex), suggests one more possible mechanism for the introjection of an illusory *"self"* onto yet another empty center stage.

Shifting our attention to the frontal lobes of the neocortex (the areas associated with an array of uniquely evolved forms of human intelligence that include language, rational judgment, planning, problem solving, and memory), we find here the very same arrangement of cortical columns stacked perpendicular to the pia mater that so plainly characterizes the structure of the visual, auditory, and somatosensory cortices. This is not conclusive evidence by any means, but it is certainly suggestive—that the frontal lobes themselves may be capable of producing some form of *mental sensorium display*. The theory of neurobioluminescence in the evolution of sensorium consciousness radically reframes all aspects of mentation—what we would normally consider the "higher" deliberative functions—and places them squarely back into the context of *sensation*.

In the orthodox Buddhist view, the mental faculties (all forms of thought, language, intellectual analysis, memory, imagination, and future projection) together comprise a sixth *sense*, also called *mind*, which in absolute terms is no more elevated in status than vision, hearing, smell, taste, or touch. The Buddhist analysis of consciousness has avoided many thorny conceptual traps by placing the mind at the very same level in the organizational hierarchy as the other five senses. With the mind's demotion to a sense modality for the perception of "things that may or may not be there," it should come as no surprise if we were to find structures of presentation neurons here upon the surfaces of the frontal lobes as well.

The strongest circumstantial evidence in support of this possibility derives from the fact that the histological (cellular) organization of the frontal lobes is nearly identical to that of the other sensory cortices, with the same six neuronal layers traversed by cortical columns—their capitals poised and ready to pixellate the surfaces of the pia mater with their electromagnetic displays. If we can set aside for a moment our natural bias toward viewing *external reality* as the only proper object of a valid *perception* and the job of the senses being to confirm and reify that external reality, we would have to concede that at least upon first glance—from the cytoarchitectural similarities—there is some evidence that the frontal lobes may in fact be the location of a "sensory cortex" for the *presentation* of *mind*.

Admittedly, the mind is much more difficult to conceptualize as a sense organ. Unlike the eyes and ears that poke out into the surrounding world, mind's "organs of perception" all face inward and are completely hidden within the skull. Although the biological substrate of mind is as "meaty" as any eyeball and as blatantly *presentational* in structure as the striate cortex, it is hard to shake the mistaken view that the physical matrix that sustains the mind should *itself* somehow be as subtle, "spiritual," and insubstantial as the ghostly thought forms, dreams, and intuitions that it is charged with reporting back to consciousness.

The mind's perceptual field is internal—ranging over an imaginal landscape composed of a combination of stored memories, unconscious or archetypal dream imagery, and objects identified through other sense modalities. In processing this sense data the mind subjects this imagery to a variety of mental operations, manipulations, categorizations and analyses—and then presents a final *hallucinatory display* in the form of a reconstructed memory, a visualized plan of action, or a daydream. These "thought form displays" typically manifest as faint, transparent "overlays" upon the visual, auditory, and somatosensory presentations. All such projections of mind, including the most rarefied intellectual thought forms, seem to possess at least *some* sensuous elements. It may be that the frontal lobes have developed their own more elusive "special effects"—using cortical columns designed to pixellate the "mindscape" in a flash of imagination—one that either momentarily interrupts, melds with, or (in the case of a deep daydream) entirely *replaces* the mundane sensorium surround.

On the other hand, the neuronal activity of the frontal lobes may turn out to be far more *computational* and less directly *presentational* in nature than the other senses. Thought forms may enter consciousness by an altogether different route—as somatosensory, auditory or visual imagery "forced" back upon the sensory cortices from the frontal lobes, but *perceived* by the lightbody in the usual manner—as momentary interruptions to the regularly scheduled programming.

FURTHER CONJECTURE

It is interesting to note that the mind, in general, can only interpose relatively *hazy* imagery upon the much more robust and clear-cut visual, auditory, and somatosensory landscape. We do our best thinking with our eyes closed in a quiet room or absorbed in scanning the neutral black and white squiggles on the pages of an engrossing book—situations in which the imagination is not crowded out by more powerful and immediate audiovisual input. Perhaps this speaks to the evolutionary recency of the human sense modality of mind. It seems reasonable to assume that such important functions as thought and imagination should be granted a clearer and more forceful presentation. Is it possible that the human prefrontal cortex has simply not yet had a long enough evolutionary gestation period to construct a more compelling and palpable noösensory display? This may represent the next evolutionary thrust in postmodern humans—a movement toward a more substantive and realistic imaginal presentation.

Placing the highly evolved human mind in the same category as the extraordinarily well adapted sense organs of other creatures (the sonar of dolphins, the radar of bats, the magnetic navigational compasses of migratory songbirds, and the ultraviolet vision of spiders, to name but a few distinguished examples) may help to underscore our kinship and basic equality with all other sentient beings. Excessive pride in the power of the human mind is an unfortunate source of our unwarranted disdain for the "lower" forms of consciousness that we imagine reside in other animal species. With this shift in perspective, we are all suddenly on a much more equal footing—for we are all, fundamentally, creatures of *pure sensation*. Our coveted "uniqueness" as a species stems from the special status that the human mind cavalierly grants to the human mind and as such represents the final *mental obstacle* to understanding the intrinsically egalitarian nature of sensorium consciousness. Our loudly trumpeted triumph becomes our downfall.

Let's take a moment to consider once again the cortical columns within the primary and secondary visual cortices, called blobs, that code for color. *These* columns are exactly analogous to the tightly compacted strands of silk yarn whose brightly colored tips combine to form the surface patterns of the Persian carpet. By logical inference I would have to postulate that the expression of color differentiation by these blobs must be made manifest to the lightbody by some clearly perceptible difference in the quality of light produced by the firings of the particular neurons that form the cortical columns that code for each color. This is, after all, the hypothesized location of the plane of presentation—from this point on no further neuronal processing is possible. Whatever distinctions in qualia are perceived, they are perceived because of real differences discernible at this juncture, where the brain's activity becomes the lightbody's experience. This could be accomplished by a modification in frequency or pattern of neuronal firing or better yet—as I would have it in my patent application's "preferred embodiment"—by an actual difference in the wavelength signatures of the electromagnetic energy produced by the neurons of different color species of blob columns. That would certainly be the most straightforward solution—having each blob be responsible for producing a single pixel of a given specral "color" at a particular location on the visual cortex "screen."

I predict that neurobiologists will soon discover obvious physical differences—perhaps specific, identifiable *pigment* molecules or distinct *refraction layers* (like the layers of karatin film coating the barbs of the feathers of the peacock's tail) within the cell walls or cell bodies of each of the various categories of presentation neurons. These would function either as simple colored filters or perhaps more complex structural optical devices, allowing a given neuron to reflect, refract, or filter the electromagnetic waves produced by its firings—thus focusing and intensifying the presentation of a particular spectral "color" of *Light*. Perhaps these distinctions in "coloration" between subclasses of presentation neurons are the result of perceptible differences in the spontaneous biophoton emissions from a palette of as yet unidentified bioluminescent proteins (such as *luciferase*, the protein that causes a firefly's tail to glow). In any case, the differentiation between the specific color qualia produced by these blobs must somehow be clearly distinguishable *here*, upon the surface of the cerebral cortex, and such lighting effects can only be accomplished by the patterned firings of the cortical neurons themselves.

The same reasoning would apply to cortical columns of all types. Columns that code for a particular edge orientation must *themselves* be arranged in that exact orientation, thus forming simultaneously firing groups of similarly orientated columns in order to "illustrate" to the lightbody the feature they have detected. Columns that code for *movement* in a particular direction must be interconnected in such a way that they are able to work in concert with one another to "mimic" in the brain the movement in the environment that the eyes have perceived. As a general principle, cortical columns that fire in *response* to particular qualia are simultaneously *producing* those qualia—either directly upon the surface of the cortex or elsewhere in other topographic maps to which their neurons project—allowing their presentations to be immediately perceptible to the lightbody.

Therefore, the evolution of any new mechanism in the sense organs that is capable of *distinguishing* a given category of sensory input (for example, a certain class of retinal cone cells in the *eye* evolving a particular sensitivity to electromagnetic wavelengths of around 440 nanometers) must be fortuitously paired with the evolution of a corresponding neuronal correlate in the *brain* capable of *presenting* that quale (the visual experience of "blueness") directly to consciousness. Only by means of such functional *pairings* of aspects of the outer and inner worlds is the sensorium itself able to evolve into a more elaborate and integrated virtual reality environment.

Parenthetically, this line of reasoning helps us better date the transition from cybernetics to sensorium consciousness in the lineup of human ancestors. When one considers for even a moment the incredible intricacy and verisimilitude of the *presentation* of the somatosensory, auditory, and visual realms of the sensorium, and one begins to count the number of *paired* evolutionary breakthroughs that must have occurred in *both* the functional operation of sense organs and the special effects presentation strategies of the brain, one must admit that this is overwhelming circumstantial evidence, bordering on proof positive, for the immense antiquity of sensorium consciousness. It would require a tremendously long evolutionary timeframe, in other words, to work out all the bugs. If the brain were simply a cybernetic system that did not need to *present* its findings (as it self-evidently *does* to *humans*), such *pairings* of innovations would not be necessary. As it stands, the notion that any of our recent prehuman ancestors—or, by extension, our more "primitive" contemporary animal cousins—would *not* experience their own versions of sensorium consciousness is for this reason alone a logical impossibility.

Now that you have become more familiar with the guiding principles of this theory, you would probably be struck in the same way I was by the color illustration I came across the other day—certainly viewing it with a keener eye than whoever wrote the accompanying copy (Fig. 27.13 D on page 744 of *Fundamental Neuroscience* [second edition], edited by Larry R. Squire et al. 2003). The computer-generated image of apparently random splotches of color displays data from a portion of the surface of V1, color-coded to show the locations of cortical columns that respond optimally to a particular angle of orientation of visual stimulus. To the side of the image is a key, in which differently colored line segments fan out, representing orientations of 0°, 30°, 60°, 90°, 120°, and 150°. Although initially the image appears to show randomly spaced, somewhat oddly shaped areas of color, if you squint a bit and concentrate upon a single color at a time, you will see that indeed the cortical columns are *themselves* clearly arranged in lines parallel to the angle of orientation they are supposed to represent! No mention is made of this anywhere in the text, undoubtedly because neurobiologists have had no reason to look here for such simple structures of *presentation*. This color illustration can only demonstrate that the cortical columns are properly *arranged* for presentation—as mentioned above, the rows would have to fire *simultaneously* in order to present the visual effect of an edge.

It stands to reason that any areas of the cerebral cortex that are *not* devoted to sensory presentation would find themselves relieved of the pressing need to maintain strict columnar organization. In any such area, cortical tissue should be able to arrange itself like a proper biocomputer, taking full advantage of its three-dimensionality to engage in unencumbered evolutionary and developmental complexification. Only those regions devoted to the perception and presentation of sensations and thoughts would limit themselves to pure columnar organization, thus betraying their well-kept secret—that their capitals form a mosaic *surface* over which something is moving and to which that something must have immediate access. Only a *two-dimensional* surface can provide this ready accessibility.

As unfathomably complex, resplendent, and spellbindingly beautiful as this "Persian carpet" may seem at first glance, such paltry praise gives only the vaguest hint of the tiniest fraction of its full magical power. The patterns and pictures displayed upon it are not fixed, but move in resonance with the shifting world. Multicolored forms pass by from side to side and top to bottom across its surface, re-creating images in brilliant *Light* of whatever objects may happen to present themselves to the eyes. And so it is for all objects of awareness in all the various sensory presentation forums of the brain. Sights, sounds, smells, tastes, sensations, and mental imagery arrive as a crowd of splendid stars upon the red carpet, decked out in their finest qualia. They are the ebullient surge of perpetual first-nighters, an audience of multitalented improvisational performance artists, who pack the Neo-Cartesian Nondual Theater of Consummate Simultaneity for another original stage production of *A Moment in a Day in the Life of No-Self*.

Contrary to what you may have read in the fairy tales, it is not the "magic flying carpet" that moves. *We* are the ones doing the flying, while the carpet stays landlocked, firmly attached to the surfaces of the brain. It is only the *sheen* that moves, a glistening mirage appearing and disappearing, as we glide over the colored sands, the pixellated tips of the cortical columns that form the vast expanse of these undulating Sahara dunes. Visualizing yourself as the lightbody, you surround the entire neocortex, embracing the gentle mother, experiencing everything—all feelings, all perceptions, and all products of the imagination—as pure sensation and pure *Light*. Such simple exercises as this may permit you, perhaps for the first time, to catch a glimpse of the way things really are.

Then what is the pia mater to the sensorium but a little Vaseline on the lens?—the portrait photographer's secret weapon and best friend of the aging beauty queen. The pia mater itself, being only three cells thick, provides minimal physical protection for the brain. Although that is certainly one of the pia's occupations, it is also moonlighting for the sensorium, providing the soft focus that blends the pixels and warms the imagery.

And just beyond the pia mater is the crystal-clear ocean of *cerebrospinal fluid*—reduced in the cramped imagination of the neurobiologist to the role of a hydraulic shock absorber. The pia mater is separated from the much thicker *dura mater*—"the hard mother" that lines the skull—by the *arachnoid mater*, the "spidery mother," whose arches of transparent, collagenous fibers create the grotto through which the cerebrospinal fluid flows. These three structures—the pia, the dura, and the arachnoid mater—are collectively referred to as the *meninges*. The fibrous filiments (called arachnoid trabeculae) have their more formidable, substantive parts up near the dura mater and only lightly touch down with delicate footprints upon the pia mater. The flimsiness of these structures would seem to be an inexplicable mechanical oversight if the watery expanse of the subarachnoid region functioned purely as a hydraulic shock absorber. But the very insubstantiality of the trabecula at its point of contact with the pia mater is actually strong corroborative evidence for its being a compromised structure, one that intends also to minimize its profile—just as the flying buttresses that support the walls of Notre Dame Cathedral are delicately arched and hollowed out to allow as much sunlight as possible to enter the magnificent stained glass windows and fill the cavernous darkness with dazzling illumination.

Placing the lightbody "outside" the brain in this way allows us to see the interface between spirit and matter clearly and distinctly, once and for all. I'm under no illusion, however, that the brain would maintain that degree of oversimplification merely to accommodate the desire of Western scientists for a neat Cartesian dualism. Now that we have established the conceptual distinction between the realm of the spirit and the realm of matter, let's look more closely to examine the incredible melding of realms that actually takes place—*Light* interpenetrating flesh—forming the friendly frontier of the cerebral cortex itself. We will be looking more closely at the brain's cytoarchitecture for evidence of structures that appear to be undergirding this staggeringly elaborate light show.

Neurons are held in place by *astroglial cells* that actually comprise 50% of the total number of cerebral cortex cells and 20% of the brain's volume. Astroglial cells have tiny cell bodies, but their long processes serve to hold the brain together in a web-like structure. They are in contact with blood vessels and help to form the blood-brain barrier. They encapsulate synapses and neuronal cell bodies and hold them in place. They allow the brain to maintain its peculiar structural form, a spongy mass riddled with holes. The brain has not employed the brick and mortar technique in its construction, as is invariably found in all other organs of the body. It would be difficult to account for this degree of whispy insubstantiality in the body's most vital organ, were the open spaces not serving a very important function.

Astroglia (meaning "star-like glue") are similar in shape to neuronal cells, in the sense that both cell types seem to be trying to "get out of the way" of something, to minimize their profile within the crystal-clear ocean of *cerebral interstitial fluid* that fills the extracellular space and in which all neurons, astroglia, and other brain cells float. This transparent fluid makes up a full 20 to 27% of the total volume of the brain. Whereas all other organs of the body are suffused with capillaries whose semipermeable membranes allow the cells of the organs direct access to the bloodstream, permitting the easy exchange of a wide variety of gases, fluids, and nutrients, the brain has developed an elaborate system, called the blood-brain barrier, specifically to minimize that sort of casual contact. It consists of special blood vessels—coated with endothelial cells forming tight junctions and surrounded by an impenetrable basement membrane—that enter and exit through tight-fitting openings in the pia mater. These blood vessels meander through the space of the brain like pneumatic tubes in an old-style office building. Designed to perform the essential metabolic functions of capillaries while taking up a bare minimum of space, these special blood vessels contain and segregate what is *inside*—the semiopaque red blood cells (with their problematically iron rich, magnetically charged hemoglobin), the straw-colored serum, and the hormones, salts, essential plasma proteins (albumin, globulins, and fibrinogen), and discolored metabolic wastes (such as urea)—from the brain's dazzlingly clear liquid atmosphere.

If you have been following this line of reasoning closely, you may be tempted to jump to the same conclusion that I did. There seems to be some rather suspicious activity going on in the cerebral cortex. For one thing, it doesn't appear as though an optimal degree of spatial compacting has been achieved by *this* biocomputer. If the neurons themselves are the be all and the end all, the "little brains" within the brain, why aren't they packed in there like sardines? Why enlist the aid of more than 100 billion astroglial cells to keep those neurons at arms distance from one another? Why this delicate filigree? So much unused real estate makes me suspect that the wedding of Cupid and Psyche might actually be happening out in the bright sunshine on the great front lawns, leaving the erratically lit mansions in the care of sleeping servants.

Crystal clear cerebrospinal fluid bathes the entire cerebral cortex from without. Crystal clear cerebral interstitial fluid permeates the entire cerebral cortex from within. Many good reasons can be cited for establishing a blood-brain barrier and a blood-cerebrospinal fluid barrier. It is presumed that these structures evolved primarily to protect the brain against microorganisms and cytotoxins and to allow the brain to have its own self-contained neuropeptide environment, unaffected by the body's peripheral neurotransmitters. The cerebrospinal fluid that fills the meninges that surround the brain is assumed to be not much more than shock absorber fluid—there principally to shield the delicate brain from trauma. But why not toughen up that delicate brain a bit? It seems odd to have gone to all that trouble, evolving such a complex system to protect a brain that has the consistency of a runny soft-boiled egg—easily "scrambled" by a sucker punch to the jaw—instead of beefing it up with a bit more material substance and draining out some of that sloshy excess fluid.

So the whole cerebral cortex—including the flooded galleries of the meningeal auditorium—takes part in the light show. The electromagnetic fields generated by the firings of those 100 billion neurons intermingle even there, largely unimpeded in this primordial ocean of *Light*. The cortical columns that organize this display are not columns of solid marble—more like strands of gossamer fiber-optic thread—opening up a three-dimensional space that appears like a roofless, sunlit Parthenon, some two to five millimeters thick. That should leave plenty of room to "flesh out" a three-dimensional world along those cortical columns that code for depth. But how *do* they manage the trick after all? Could that mysterious *quale*—depth of field—that gives this whole presentation its final artful master strokes, actually be compressed into these few millimeters of cortex?

Perhaps the deep convolutions of the brain have evolved for more than their primary purpose of maximizing cortical surface area. The convolutions of the occipital, parietal, and temporal lobes may long ago have been enlisted to play a more direct and active role in producing the palpable illusion of three-dimensional space in the brains of higher animals.

Somehow from inside a head that fits snuggly within a baseball cap, we project our own personal planetarium high into the evening sky—far, far above our heads, but nowhere near infinity—where soaring birds and high clouds and moon and planets and constellations commingle upon the same curved surface of the inverted teacup—the ultimate backdrop—the sphere of fixed stars that circumscribes the farthest reach of the imagination.

FURTHER CONJECTURE

A SUMMARY OF CHAPTER FOURTEEN

The theory of neurobioluminescence in the evolution of sensorium consciousness provides a new framework for exploring the structures and functions of the sensuous brain. This chapter illustrates a new methodology for analyzing its cytoarchitectural features and neurological processes, one that allows "felt experience" and introspection to be employed as legitimate research tools.

When we perceive an object in the world, what we are actually perceiving is a luminous bustle of neuronal activity within the brain itself. Each time we experience a particular physical sensation, a sound, a thought, or a visual form, we have an opportunity to become mindful of the fact that a precisely corresponding neuronal analogue must be executing its presentation at that very moment somewhere upon the surface of our cerebral cortex. Practicing in this way, the fully conscious embodiment of the bioluminescent brain becomes a powerful, self-reflective meditation upon the genuine and palpable interconnectedness of the body, the mind, and the external world—no longer a hypothetical or sentimental spiritual construct.

The first thought experiment proposes that the ocular dominance columns in the primary visual cortex may be directly responsible for the presentation of three-dimensional space. The second proposal suggests that the layered structure of the lateral geniculate nuclei may also facilitate our experience of visual dimensionality. The third thought experiment elaborates upon the Buddhist conception that the mental faculties of analytic thought, imagination, memory, and conjecture operate as a sixth sense, in function largely indistinguishable from the other five.

The chapter concludes with a physical description of the overall structure of the cerebral cortex, the cerebral spinal fluid filled meninges, the cerebral blood vessels, and the blood-brain barrier, arguing for a complete reexamination of the ultimate purpose of this cluster of features in light of this new theoretical model.

CHAPTER FIFTEEN

THE STREAM OF CONSCIOUSNESS

Since you are evidently reading this book, I can presume that you are awake—perhaps you are in your living room, comfortably ensconced in an easy chair—or propped up on your elbows, stretched out on a beach towel in the sun. I want to make sure that you are *wide* awake before we proceed, so look around for a moment. Take it in. Where are you? What do you see? What do you hear? How do you feel? No need to answer out loud. Just experience yourself fully in this moment.

Whether you know it or not, what you are in this moment is a being of *Light*—receiving images of the world in tangible forms composed of your own substance. The light that you see is the *Light* that you are. The trees, the grass, and the flowers that may appear to surround you, the lilting bird songs you hear, the breeze you feel on your skin—it's all *you* and it's all *Light*—and it's *not* a metaphor! You are, quite literally, the *Light Within* that *itself* has been formed into a three-dimensional representation of your *self* and the surrounding world. You are not something separate that is *touching* your experience. You are a shape-shifting energetic field headquartered within your skull—and extending, quite possibly well beyond. You are a

THE STREAM OF CONSCIOUSNESS

protean conscious environment, simultaneously subject and object of your experience.

Beneath you, sustaining you, informing you, is a mechanism of great subtlety—a sparkling chain mail platform, ever so "intelligently" designed. I hate to call it a biocomputer, but where else to begin? It's a living organism—organizing and reorganizing itself in every moment—a myriad of axons and dendrites incessantly forming new connections and letting go of others—never the same brain twice.

The flurry of "information" passing between and among these hundred billion neurons with their 164 trillion synaptic connections is beyond comprehension—or perhaps more accurately—*beneath* comprehension. The intense "computational" activity of the brain is completely removed from what we experience as thought and sensation. No amount of this sort of "information processing" ever adds up to anything resembling consciousness. Neither is this "information" meaningful to the neurons themselves. They are simply following the most rudimentary nonconscious algorithm: so much synaptic input within such and such a time frame—fire! Recover. Wait for another signal. Do it again. The same basic instructions that appear on your shampoo bottle: Lather, rinse, repeat...

At any given moment—when they are not diligently reengineering themselves into an even more adaptive, sensorium-creating biocomputer program, that is—neurons in the brain are actually behaving like a passive medium. The firing of one neuron will have an entirely predictable inhibitory or excitatory effect, or some combination of both, upon the dendrites of the succeeding neurons to which its axons are synaptically connected. Like the attractive and repulsive electromagnetic forces between water molecules (within the greater context of the gravitational field) that determine how pressure waves and surface waves propagate through a body of water, these inhibitory and excitatory connections also propagate waves of so-called information. There is nothing "intelligent" about a neuron's decision to fire or not to fire. It does not "decide," based upon careful consideration of the information it has received from its neighbors, whether or not to pass this rumor on down the block. Consequently, there is no information in a firing neuron other than the self-evident fact of its firing. Whenever a neuron fires, it is always giving the exact same message to the neurons to which it is connected, and the message is this: "I have just received a sum of inhibitory and excitatory signals from the other neurons to which I am connected that happened to exceed my current excitation threshold, and I wanted to let you know about it." End of story. No neuron is saying anything intelligible, such as, "I was just informed that our sector of the retina is now seeing a patch of red light. Pass it on." And most certainly no neuron or group of neurons ever experiences the pleasant shock of recognition and exclaims, "Grandmother! It's you!"

Instead, what does happen when your grandmother's memory is evoked is that an immense group of presentation neurons is prompted to display an actual electromagnetic image of your grandmother's face; a brief slideshow of a few stereotypical memories of grandmother pruning roses in her garden or baking cookies in her kitchen is quickly displayed in your visual cortex, and the word "grandmother" is elicited from memory and quietly spoken into your auditory cortex. An unimaginably complex sequence of neuronal firings—a cascade initiated by a very tiny trigger (perhaps, theoretically, by the firing of a single "grandmother neuron")—has unwittingly produced this flood of electromagnetic sensorium imagery, all of which is entirely meaningless and in fact unknowable to any aspect of your physical brain. You are the only one home to receive these messages, because only within the *Light* could such visions be seen and sounds be heard. No possible analysis of the firing patterns of these neurons would reveal anything "grandmotherly" about any of the signals being sent between neurons. The grandmotherliness exists only in the presentation images themselves. Like a giant portrait of Mao Tse-tung formed by the communal reflex of a stadium full of card-carrying Communists, grandmother's image was similarly lifted up above the neuronal hoi polloi. How her image—and not Mao's—came to be selected by attention is the subject of a later chapter.

Brains do develop and learn, of course. Neurons readjust their synaptic connections in response to heightened activity, increasing the number of dendritic synaptic connections to those neurons that provide regular, patterned input and releasing their connections from those that do not. This process is based upon algorithms encoded within a neuron's genes and is also influenced by the transient presence of specific neuropeptides. The cumulative effect of this ongoing reorganization is staggering—it is able to transform a baby banging a

THE STREAM OF CONSCIOUSNESS

spoon upon a cooking pot into a concert pianist within the course of just a few decades.

The point I am trying to make is this: that beneath these crucial processes that lead to the brain's complexification *in the long run* is an even more basic property that allows the brain to operate smoothly and predictably *in the moment*. A resonance neuron does its best work when it adds absolutely nothing to the ongoing information exchange. It is being asked only to behave absolutely predictably according to its own individualized set of rules of engagement—the particulars of which need be known only to that individual neuron. If every neuron in the sequence follows this edict of noninterference, the brain is able to function in its fundamental capacity—as a complex "chaotic" resonator of neuronal firing patterns—and is able to create (as a byproduct, from its perspective) the finely detailed electromagnetic field that we experience as the sensorium.

What the physical brain is concerned with (the computational activity of its intricate synaptic connections), the lightbody is entirely unconscious of. Conversely, what the physical brain is "unconscious" of even having produced (its epiphenomenal electromagnetic field), the lightbody experiences as the very stuff of consciousness.

This is a key insight that will help to clarify the basic premise of the theory of neurobioluminescence in the evolution of sensorium consciousness. Neurons have been arranged by evolution and development into structures that appear to be intelligently processing information—and these structures have naturally been likened to biocomputer circuitry. Neurobiologists are very fond of referring to the brain as an "intelligent machine," and inferring that its intelligence (and therefore its awareness) is somehow a result of all of that interneuronal computational activity. If neurobiologists insist upon the use of this type of imagery, they will continue to impute "intelligence" and "purpose" where there is none. In actuality the brain has been carefully sculpted by evolutionary design and by life experience to be, within the context of a given moment, an intricately structured, but ultimately passive instrument—a medium, as it were—for the propagation of waves of nerve impulses. No thought is involved in the entire process; no decisions are ever made by neurons whether or not to fire in a given situation—not at the level of the neuron, nor at the level of more complex neuronal structures, not even at the highest levels of organization of the brain itself. The specific criteria for firing (within a given transient neuropeptide environment) are preprogrammed into each neuron. Although differences in cell types and configurations make it impossible to predict the cumulative global outcome, nonetheless, at every level of organization, firing patterns are determined by nonconscious biophysical algorithms. It is a sloshing back-and-forth of nerve impulses, as precisely mathematical in its propagation and as ultimately chaotic in its development as any other complex vibrational medium.

To illustrate this, let's return to an image from chapter eight. Once again comparing the neuronal substructure of the sensorium to the highly refined material components of a well-crafted violin, we can easily see how the specific vibrational properties of the foundational substance—wood—though entirely distinct from those of the final medium of presentation—air—are clearly the sole material cause of the complex propagated sound waves within the air that we enjoy as music. A master craftsman has carefully sculpted the back and belly of the wooden sound box so that *it* may, as an epiphenomenon from its perspective, propagate vibrational sound waves through the air of a particularly beautiful harmonic structure. The instrument maker cannot directly carve the *air* into the musical forms he desires to hear. As a clever alternative, he has learned to carve the wood into the precisely appropriate form so that *it* can subsequently "carve the air." The air cannot produce its own music without the aid of the violin (just as the exquisite form of the lightbody cannot exist as such without the patterned firings of neurons in the brain). The wood of the violin is completely ignorant of the music it is producing in the concert hall (just as the physical brain is ignorant of the lightbody it is producing). The fibers that comprise the wood are responding as a passive medium, stretching and compressing according to the physical laws that govern their movement, their momentum, and their exchanges of kinetic energy (just as the great web of neurons in the brain forms a passive medium of so-called information). Human craftsmanship has sculpted the raw materials of maple and spruce into a magnificent chaotic resonator, capable of transforming the subtle vibrational patterns of string and wood into the equally subtle vibrational patterns of an entirely different medium—the open air of the concert hall (just as the process of evolution has carved the raw meat of the brain into a magnificent neural resonance chamber, capable of transforming streams of sense data into the *Light* of sensorium consciousness).

In addition to its many other practical occupations, the brain may be viewed as the organ that funnels linear streams of sense data into discrete cortical areas designed to function as highly specialized chaotic resonators—chaotic having a precise mathematical meaning here. Analyses of the firing patterns of neurons reveal telltale signs of a vast amount of reentrant looping. Mathematical chaos emerges because neurons are so richly interconnected that the output from a given neuron eventually becomes a part of its own input in a later iteration. There are so many outputs that are reintroduced as inputs that no specific outcome can ever be predicted—but predictable patterns nonetheless emerge.

These patterns of firings exhibit a certain overall form and relative stability and are known as chaotic attractors. This type of resonator has the ability to shift between a large number of relatively stable states, such global changes often being initiated by a very small external input. Sense data is one of the sources of destabilizing input capable of shifting the patterns of resonance from one chaotic attractor to another. The neural resonance chamber as a whole is chaotic, being made up of many other smaller chaotic neuronal resonators, themselves comprised of tiny neuronal oscillators, all richly interconnected.

The concept of chaos takes a bit of getting used to. Having read a number of books on chaos theory and having spent a good deal of time wandering through online "fractal art" galleries absorbed in their amazingly intricate, psychedelic video images—I've begun to acquire a feel for chaos and see within it the likeliest mechanistic explanation for how the brain is able so readily to create, sustain, and destroy the sequence of forms we perceive in the sensorium. This modern incarnation is not the same creature as the unbridled, unknowable Chaos of Greek mythology. A "chaotic system" is in fact the perfect bridge between the mathematically rational (involving fractional relationships between whole numbers that form, for example, pleasing musical intervals and chords) and the irrational (transitional incoherence, experienced as noise or as nothing in particular). Chaos gives an adaptive quality to all complex, organic resonators. It is what allows them to promote and sustain a given waveform in one moment and then just as quickly reestablish themselves into a new vibrational mode when presented with a different input waveform.

Therefore, referring to a violin as a chaotic resonator should not be taken as an insult to the violinmaker; for it does not imply that the instrument is only capable of making incoherent noise. It means that the violin is able to receive and respond to a wide variety of vibrational patterns. After an initial moment of settling in, perceived as the characteristic "attack" at the beginning of the draw of a down-bow, a steady pitch is established upon the vibrating string and translated through the bridge to the arched top of the violin. A chaotic moment follows as a unique configuration of nodes and antinodes is located and engaged in the vibrating sound box—the one configuration capable of sustaining that particular pitch. This is the reason why the violin has such a voluptuous, curvilinear form—often compared to a woman's body. There need to be larger, thicker areas in the belly for the full range of low notes to expand into as well as finely tapered corners to initiate the resonance of high notes and upper harmonics. A fine violin is a complex organic resonator designed to reverberate sympathetically across the full spectrum of vibrational frequencies. Musicians would describe such an instrument as being responsive. It exhibits alacrity, like the alacrity of mind.

In the same way, the brain creates structured images of presentation composed entirely of electromagnetism—sustained not by any vibrational resonance of the electromagnetic field itself, but by resonance patterns formed through reentrant sequential firings of large-scale neuronal structures. From the neurons' perspective—if they had one—the electromagnetic images that they are creating are pure epiphenomena, of which they are not aware and which have no immediate effect upon their own patterns of firing. The electromagnetic images exist as if in a separate "space" discernible and meaningful only to the lightbody, just as the violin music exists in the concert hall, not in the chips of wood that make up the violin. This is a very strange way of thinking about thought, but until the brain and the lightbody it sustains can be conceptualized entirely independently of one another, a real understanding of the theory of neurobioluminescence in the evolution of sensorium consciousness will remain forever elusive.

THE STREAM OF CONSCIOUSNESS

Normally we think of a medium as an unencumbered expanse, like the ocean or the atmosphere, through which well ordered waves make their predictable passage. When thinking about the mind wending its course through the brain, it is more useful to imagine the medium as if it were a mountain stream. The water here appears to be a different element altogether than that which fills the ocean depths, but it is not. It flows along its course, obeying the same hydrodynamic laws that govern the movements of the oceans. It begins as a trickle emerging from snowmelt high in an alpine glen. All along the way it is jostled and splashed, broken and rejoined by the rounded boulders over which it tumbles. As it makes its way down the steep valleys of the high mountains it is joined by more clear water seeping in from the saturated slopes. It fills up every pool, bounding over rocks, cascading over waterfalls, growing ever more powerful, carving out its own path, uprooting trees and stubborn boulders, becoming eventually a mighty river flowing inexorably to the sea. And yet at every point along the way you will see the same predictable structures: standing waves, well-defined areas of turbulence, and other periodic fluctuations in the current. Although it is constrained to flow through a complex path of pools, spouts, and waterfalls—the water itself always behaves as a passive medium for the waves that travel within it.

Just so, your mind begins as a flash of *Light* in the brainstem, born into existence in a moment of bewilderment and unknowing. Brought into being within this dark womb, it is swallowed upward by luminous contractions—pitch blackness behind, growing light ahead. This spark of *Light* is conscious and from the very beginning separate from the rocks and boulders, the neuronal structures over which it travels on its way. Drawn into the growing illusion by an irresistible invitation, the faithful pilgrim follows the ancient, well-lit path. He is always the sparkling new-comer, like the title character of Voltaire's *Candide*, credulous and ever attentive to the instructions of Master Pangloss:

> "It is demonstrable," said he, "that things cannot be otherwise than as they are; for as all things have been created for some end, they must necessarily be created for the best end. Observe, for instance, the nose is formed for spectacles, therefore we wear spectacles. The legs are

visibly designed for stockings, accordingly we wear stockings. Stones were made to be hewn and to construct castles, therefore My Lord has a magnificent castle; for the greatest baron in the province ought to be the best lodged. Swine were intended to be eaten, therefore we eat pork all the year round: and they, who assert that everything is right, do not express themselves correctly; they should say that everything is best."

The lure of *Light* draws the pilgrim onward into growth, development, becoming—finally joining up with the brilliant, resonant community of mind that welcomes him into the fold, enfolding his fresh newness as a prized commodity—the eye of wonder just now emerging from the sacred fountain.

A person resuscitated from clinical death may report having ascended through a gloriously radiant tunnel and having been urged by the voices of friendly, disembodied guides to "go toward the light." He may recall having been presented (by a guardian angel or some less farcical proxy of Master Pangloss) with a timeless, instantaneous life review—"My life flashed before my eyes," the person may say. But *where are you*, I ask you, when your failing brain can scarcely propel you beyond the bare awareness of coma, when you are no longer a ferocious waterspout but a mere dimple on the surface of the great ocean of *Light*? Beyond the neocortex or before it? Do you not hear, in these reports, a curious coincidence with the Book of Microgenesis?

The rational aspect of mind cannot conceive that it reduces, not to the boulders, not to the water, not even to the splash, but to the rainbow caught in the mist that hovers above the splash. We would hope to have ourselves be more substantial—the water at least or the rocks—the tumble of surfaces that halt and pass the water on its course. And ego, that great lunkhead, would counsel us to hunker down into a tight, muscular pose and insist upon our identity with the body and its solidity. Thank God he does not run the show and is himself more likely the illusion, the soon forgotten dream of the real rainbow spirit that dances upon the misty waterfall. In reality, all of these are connected in a oneness. The body is the mountain—the brain is the deep valley over which the water flows. You are the *Light*, pure and simple, that dances upon the stream.

THE STREAM OF CONSCIOUSNESS

A SUMMARY OF CHAPTER FIFTEEN

In previous chapters, sensorium consciousness was presented in its nondual aspect—as an intimate weaving of body and spirit. Here, the opposite tack is taken—the two realms are split asunder—in order to contrast their separate roles and responsibilities. The first distinction that is made is that you are the Light—the energetic field—that perceives, experiences, and is vitally aware. What you are aware of are the electromagnetic presentations produced by the neuronal activity within the sensory cortices. However, these interneuronal communications are not transmitting the "information" that somehow adds up to your awareness or your intelligence. Neurons are simply following strict computational algorithms when they fire. The brain in its entirety and the individual neurons that compose it are presumed in this theoretical framework to be completely devoid of intelligence, sensation, or feeling. They are simply the physical substrate that permits these qualities to blossom in the energetic realm of mind.

Within the context of a given moment, the brain behaves like a passive medium for the propagation of nerve impulses. In this way it is able to fulfill its function as a complex, chaotic resonator—a Light-violin—that accurately reflects the images and sounds of the world brought in through the senses. Combining this imagery into one coherent whole is the job of the sensorium. Continuously refreshed, awareness reenters this sensuous display in an endless succession of conscious microgenetic moments.

The near-death experience may be an intimation not of what comes next, after the final moment of this life, but rather of what comes before and during each and every microgenetic moment, as recalled by a person whose dying brain could no longer obscure or fully erase this preconscious developmental phase. The "trustworthy" voices of fixed neurobiological archetypes of our "ancestors": of mother and father, siblings, friends, and spiritual guides—all routinely appear to surround us in each instant of becoming, urging us to "go toward the Light." Perhaps it is not the road to Heaven or the Godhead but the bright path toward the neocortex that the voices are referring to!

CHAPTER SIXTEEN

CHILD'S PLAY

Descartes may have had a few problems with his epistemology, but he did invent a damn fine coordinate system. We are going to employ it here to map out our own personal *cube of consciousness*. This is a conceptual model that should help us better understand the three interrelated dimensions in the development of sensorium consciousness. We will be using the three-dimensional version of the Cartesian coordinate system, with an "x", a "y", and a "z" axis.

Imagine three perpendicular lines in space (the axes) all intersecting at right angles to one another at a single point (the *origin*). Place a mental cube with one of its vertices (corners) at the origin (whose coordinates are (0, 0, 0)) and the three edges of the cube that emanate from that vertex lined up on the positive halves of the three axes, the edges being segments whose other endpoints are at coordinates (1, 0, 0), (0, 1, 0,), and (0, 0, 1). We will confine ourselves to this one octant (eighth section of space) in which all three variables are nonnegative (positive or zero). We'll be using a left-handed coordinate system, by the way, which means that the positive direction on the x-axis points to the right, with the positive direction on the y-axis pointing down

CHILD'S PLAY

and to the left (appearing to poke out into imaginary three-space at an angle of approximately 30° below the horizontal) and the positive direction on the z-axis pointing straight up from the origin. Each of the three dimensions will be measuring a very different timeframe, but in this model we will consider the length of each edge of the cube to be one *cubit*. (... *right*... What's a cubit?)

Along the x-axis we'll plot the *evolutionary* history of your body-mind up to and including this current human incarnation, representing a progression of more and more complex forms of consciousness. The graph begins with the first primitive manifestation of sensorium consciousness (whenever in prehistory you might imagine that event to have taken place) at the origin and ends with your own fine psychophysical specimen of a self one cubit to the right, at the lower, right, back vertex of the cube. This dimension will be measured in millions of years, and all along this edge is a parade of ancestors—one selected from each generation—those barely distinguishable from you (a parent, a grandparent, and a great-grandparent) standing next to you on the extreme right and those of earlier species in your lineage slithering off to the far left.

The y-axis angles down and to the left in our mental picture, and this edge of the cube indicates your *developmental* progress from "quickening," or ensoulment, to the present—once again beginning at the origin with whatever point in the course of gestation you (or your religion) might imagine that event to have taken place. This dimension is measured in years and takes you on through birth, infancy, childhood, adolescence, and adulthood, right up to the remarkably highly developed individual that you are at this very moment. Down toward the far left we will place the more recent portraits of you. Up toward the origin, just to the right of your elementary school photos and your earliest baby pictures you'll find some ultrasound images that you might suspect had been accidentally switched for those of a monkey, a lizard, or a fish—but they indeed are pictures of you, as the very young fetus that you once were.

The x-axis maps the *phylogeny* of your own human body-mind from the first prehuman ancestor endowed with sensorium consciousness. The y-axis maps your *ontogeny*, the development of your particular body-mind from your earliest conscious moment to the present. The z-axis—which shoots straight up from the origin and measures the height of the cube—maps *microgeny* and so is a bit more difficult to conceptualize. According to Jason Brown's theory of microgenesis, you are in every moment retracing similar ancient pathways, revisiting neuronal structures representative of each stage of evolution and fetal development. New iterations of you are actually darting up the right, front edge of the cube at an alarming rate—perhaps ten times per second. However, we will be using the z-axis to indicate your prevailing mind state—whatever plane of consciousness you might have achieved and are currently enjoying. The bottom face of the cube represents the plane of dreamless sleep, which is actually experienced as pure wakefulness devoid of subject, object, time, or space. Presumably you were floating in that primordial plane of consciousness just last night, a number of times, on into the early morning hours. With the onset of each REM sleep cycle, you spent some time in dream consciousness, occasionally waking into a moment or two of groggy, initial object awareness, and now, thanks perhaps to a strong cup of Italian roast, you are experiencing full-fledged analytic perceptions and wide awake sensorium consciousness.

You are in fact, at this very moment in vector space, inhabiting the uppermost, rightmost, forwardmost vertex of the cube, at the farthest possible distance you can be from the origin, yet at every moment emerging from that same lowly plane. As an august member of the human species, you are at the extreme far right in your evolutionary progress (but hopefully not in your political beliefs). As a full-grown, well-educated, worldly-wise adult, you have stepped forward to embody the most mature expression of your mental and emotional development. To top it all off, you have risen out of your slumber and are now wide awake, sitting pretty—high atop the cube of consciousness. Congratulations... you made it.

If you turn around, you'll see that you are sitting upon a mountain that must eventually slope downward in all directions within your cutaway cube, the particular shape of the terrain determined by your own degree of generosity in attributing full or nearly full consciousness to our earlier ancestors—but with the lowest point being, by definition, at the origin. If you were to draw topological lines to indicate planes of equal elevation, you would clearly see how these vastly different beings in your ancestral lineage at different stages of development may at certain times in their diurnal cycle be said to share a relative equality of conscious experience. There are microgenetic moments that are almost indistinguishable from levels of ontogenetic development, and stages of phylogenetic evolution.

This model is intended to drive home the point that the same basic mechanism is at the core and origin of every moment of experience—of every conscious life form at every level of development and every stage of evolution. Consciousness is something that emerges from the core and attains to a certain degree of complexification. There are many levels at which consciousness may complete its particular trajectory—all are valid and all are equally real to the individual sentient beings experiencing them.

The mechanism is a *Light*-pump within the brainstem that constantly issues forth new iterations of a proto-self every tenth of a second. It propels each growing lightbody through a series of shape-shifting experiences—all laid out in a predictable sequence—hurling this newly emerging sentient being into whatever "maps" of presentation neurons have so far evolved in the sensorium of that particular organism. Before we can go any further into the investigation of how and why the brain maintains this connection with its ancient roots, we will need to have a better understanding of the specific processes involved in microgenesis.

What follows is an outline of the stages of microgenesis as described by Jason W. Brown in the fifth chapter of his book, *The Self-Embodying Mind*. Note that he does not conceptualize this process as the development of a conscious, electromagnetic entity, as I do, but instead writes simply of the *self*, whose foundational substance is not specified. In both cases, the development of this emerging *self* depends upon the generation of a specific sequence of neuronal firings, but in Brown's formulation these neuronal firings are presumably *informational* rather than *presentational* in nature. As I have mentioned before, his writing, although complex, is quite lyrical—and evocative of the very movement of the luminous mind it describes.

> *The microgenetic account of perception entails a series of moments in the developing object representation leading from the upper brainstem to limbic and neocortical structures toward exteriorization and featural modeling at the level of visual cortex. Sensory or physical*

> *input at each stage constrains the developing configuration to model the external object. The perceptual series is autonomous; layers in the final object are intrinsic mental constructs sharply demarcated from the sensory inputs through which they are defined.*
>
> *These layers, separate modes of existence, are successive planes in the same mental space. The percept begins and ends in mind. Similarly, the different types of consciousness and the varied expressions of the self linked to the different conscious states are embedded in a mental space laid down by the developing object. At least four planes of consciousness can be distinguished, corresponding to the described perceptual series.*

The first plane of consciousness he calls *pure wakefulness*, which is a kind of vigilance that is not focused on any object. It is the earliest stage in the development of consciousness, a state of dreamless sleep, similar to a coma. The *self* has not yet been formed—for there is nothing yet for it to contemplate. "The stage represents the earliest appearance of a preobject coextensive with the body and the immediate body surround in an extended somatic space field."

In the next phase, *dream consciousness*, sensation provides a preliminary representation in the upper brainstem that is then passed on to the limbic and temporal lobes. Here the object is further elaborated as personal memories and dreamwork shape the growing object. This is a dreamy state of hallucination, in which the *self* is passive, and the nature of space shifts and changes—being sometimes intrapersonal and sometimes extrapersonal. Emotional affect pervades the *self* and object, which are part of a single undifferentiated whole. It is unclear what is inside and outside the *self*.

Brown writes:

> *Inner and outer are indistinctly divided; there is a single medium that is part of the hallucinatory content. Space is volumetric, egocentric, and dependent on the viewer. Space is also a kind of object. It has a tangible, perceptible quality and undergoes distortion. The boundaries between image and space are unclear so images*

are also distorted. In addition to spatial distortions there are conceptual derailments. The similarity of shape or overall configuration as a nexus between the real object and the dream image (e.g., knife [indicating] penis) owes to prior sensory information at the upper brain stem constraining the developing object. Although relations of shape predominate, dream images tend to represent the meaning of the object rather than its form. The content of the image is determined by conceptual, symbolic, or experiential relations between the object-to-be, only some physical parameters of which have influenced the object formation at this point, and the preparatory image of that object passing through the dreamwork. The dreamer "sees" the model that has so far evolved, a model that in waking perception is derived to an exteriorized object in which the meaning is buried in the representation of object form.

Of particular interest is the importance of *shape* in the development of dream objects. It indicates to me that already at this early stage, three-dimensional space is being created by the brain, as the lightbody formulates its experience within its own electromagnetic field, which even here is produced by the firing of neurons. In the dreamwork, as an object forms, its shape appears to be known first; only later is its actual identity chosen from among competing possibilities retrieved from memory. I would understand this to mean that the three-dimensional form is experienced first by consciousness because the object itself is *composed of* the very stuff of consciousness, which always "knows" its own shape, a priori, in this most fundamental form of "proprioception." In other words, the primacy of object shape in dream consciousness is perhaps an indication that consciousness itself "fills out" these shapes with its own substance, making for a most intimate initial contact.

In the next stage, object awareness, the dream images "solidify" into more lifelike objects. The surrounding space becomes more realistic, though it is limited to the immediate environment surrounding the body. Brown postulates that this is the type of consciousness experienced by young children and also perhaps by some nonhuman animals. The self-concept is present at this stage, the result of a clearer distinction between *self* and object space. This *self* recedes into the background in the next stage as objects and object space gain more solidity.

The final stage in the microgenetic sequence, *analytic perception*, is the state in which the separation of the *self* and the world has been fully achieved. Space extends to infinity; objects are understood to exist in a public space, shared with other viewers. Objects exhibit fine featural detail, are autonomous, and are indifferent to the imaginings of the observer's mind.

> *The self needs objects that are independent; only through the representation of an external world can mind elaborate a feeling of agency that is not embedded in the world. The world has to be sought after and extracted from mind. The self is a kind of deposition bypassed in the object formation marveling at a world of its creation.*

The ramifications of the theory of microgenesis are staggering. In terms of the evolution of consciousness, this theory suggests that the human mind is simply an elaboration—an arborization or complexification—of a much more primitive form of consciousness and that the *self* is in constant contact with functional remnants of all "earlier" stages in its evolution. His theory would seem to support, by extension, the view that the "*Light*-pump" in the brainstem must have been in operation in the same location ever since it first opened up for business in our earliest conscious ancestor.

This theory has also greatly influenced my thinking regarding the growth and development of the brain during gestation. I now believe that there is something much more than the simple, passive execution of a genetically coded program occurring in the brain of a developing fetus. As soon as this *Light*-pump is up and running in the fetal brainstem, it begins producing a continuous fountain of momentarily existing conscious beings. Their investigations of the current state of the developing brain through excursions from within its associated lightbody are what actually form many of the fine-grained structures of the brain itself.

The human genome is a vast set of instructions for the production of the proteins that form and operate the human body. Genes are responsible for orchestrating the growth and development of all types of cell tissue, including the wide variety of neurons and ancillary cells of the brain. The genetic code provides the blueprint for constructing each of the large-scale neuronal assemblies and causing them to be laid down in the proper sequence, adding on more recently evolved structures once the archaic structures are in place. The complexity of the orchestration of biological processes that results in the development of even a much simpler organ, such as a lung or a kidney, is almost unfathomable. However, to imagine that the construction of a functional, conscious brain with over a hundred trillion synaptic connections is possible simply by the mechanical playing out of the genetic code defies logic altogether. Pursuing another chain of logic reveals a more plausible explanation for how such a surpassingly subtle and complex brain can indeed be created—from the inside out.

The genetic code is also what determines how every variety of neuron is to *respond* when met with input from other neurons and from the various signals of ambient neuropeptides. These chemical and electrical communications affect the growth and development of each neuron, coaxing it into creating more dendritic connections with those neurons with which it has regular, patterned commerce and dissolving other connections to neurons with which it is no longer in sync. The complex rules of engagement, though genetically constrained and therefore entirely predictable at the level of the single neuron, are so extremely sensitive to transient brain states that involve the chemical secretions and patterns of activity of millions of other neurons—that brain development soon becomes absolutely impossible to orchestrate or conduct with the requisite level of artful finesse from any central podium, regardless how intricate the score may be or how competent the mechanical conductor that reads the genetic code.

Fortunately, there is another organizing principle at play from very early on in fetal development. At a certain stage, most likely after the "reptilian" brain and other primitive structures have been formed, the brainstem begins to pump *consciousness* into the next higher layers of developing brain tissue—consciousness that eventually penetrates into the limbic system and beyond as these newly emerging structures fall into place. As you may recall, consciousness is simply an impulse of awareness—a bewildered, inchoate, curious, burst of electromagnetism, a proto-being of *Light*, unsure initially of just *what* it is becoming. Its awareness is what prompts it to investigate and in so doing to formulate and refine the specific neuronal structures as they appear—as it eagerly examines the intriguing possibilities at the outer edges of its emerging world. During this period, it is not captivated by what the eyes can see or what the ears can hear, but by the very structures of the developing nervous system and the sensory cortices themselves.

After the child is born she will continue to grow her brain in the very same way, by means of this innate curiosity and penchant for investigation. At this later stage, development forms a continuum with learning—insofar as her experiences now involve contact with tangible sense objects and interactions with other conscious beings. Although newborn infants spend a great deal of time in an apparently "unconscious" state—sleeping an average of sixteen hours per day—it is easy to imagine that they are thoroughly engaged in the further development of their brains even as they sleep.

Our own need to retreat nightly into the various phases of the sleep cycle actually corroborates this point. Since awareness itself was the primary guiding force (apart from the genes) responsible for the initial creation and refinement of all of the phylogenetic and ontogenetic planes of consciousness that can be sustained within the brain, this same awareness must therefore be enlisted back into service each night to participate directly in the subsequent maintenance of these various planes of consciousness. All higher animals, including our own distant ancestors, regularly put themselves at great risk of death or physical harm by entering this extended state of "unconscious" paralysis. If this were simply to perform the routine maintenance tasks of cellular repair and regeneration, the body would certainly have evolved some means of accomplishing these ends without putting itself so directly in harm's way. Awareness evidently needs to be present on site to do the housekeeping, to file away the learning after a dreamtime review of the day's conscious and subconscious events and to process any emotional fallout. It needs time to reengage and reorganize its archetypal structures and their neurobiological correlates. This it cannot do while on duty in the real world, in its mad rushes from the brainstem to the cerebral cortex. It must return in stages to the dreamtime and beneath the dreamtime, to engage effectively with the associated precortical structures. It needs to be "awake" and active within those very realms of existence that it was instrumental in creating—in an earlier period of its present life and perhaps as well in an earlier era of its evolutionary history.

The same inquisitive, playful nature that we associate with young children and animals is also at play within the nervous system of the unborn child. Prior to the completion of construction of coherent sensory maps, these conscious impulses are investigating other aspects of the brain and central and peripheral nervous systems of their developing body, from the core outward, up to and including whatever may be the most recently laid down stratum of neuronal tissue.

We might be inclined to imagine that a fetus lives in an impoverished sensory environment while floating in the silent darkness of its mother's womb. However, when we consider what its awesome task is during this period of gestation, we can better understand how it must be thoroughly engaged by this world of mysterious and intriguing nerve impulses with their pathways of *Light* branching out in all directions. This is the time in which consciousness learns how to operate within the brain and recognize its cytoarchitectural features, how to navigate the pathways of the lightbody and maintain its proper configuration. This is step-by-step learning at the growing edge of awareness, an expansion outward from the familiar routes that have been successfully navigated hundreds of thousands of times before.

From a Buddhist perspective, this would be understood to be a reenactment of a process that the soul has undergone countless times before. Whether or not such a long history of successful reincarnations is useful to the newly emerging being, who can say? Perhaps each pulse of conscious awareness is completely naïve, but nonetheless so similar in its response patterns to all other pulses of conscious awareness that it can act as a predictably flowing, inquisitive medium to ferret out the light-tunnels of the expanding nervous system.

At every moment of gestation, just as in every moment of life, a biological system must be *viable*. This is a self-evident fact, but still an interesting and informative contemplation. A developing biological system cannot jump from one state to another without going through the complete sequence of intermediate stages, each of which must be for a given period of time fully functional and fully viable. For example, the blood that flows through the heart and lungs must also flow directly past every cell in the body through a complex system of arteries, capillaries and veins (with the exception, of course, of the cells of the brain, which are nourished and oxygenated by special blood vessels that form part of the blood-brain barrier). No evolutionary or developmental improvements or complexifications of the body are permissible if they involve living tissue removing itself at any time from contact with the circulatory system.

Likewise, consciousness—the very "lifeblood" of the lightbody that spouts from the *Light*-pump—must during its entire gestation period and beyond have unfettered access to all regions of the brain in which it functions. Consciousness *began* in an early ancestral life form as an impulse from a *Light*-pump in a brainstem. Consciousness *begins* in a particular fetus the moment that *Light*-pump once again becomes operative. In each moment, conscious awareness springs forth from that same primogenial source. From there it must be granted entrée into all of the cerebral environments in which it operates.

The same meningeal layers and crystal-clear cerebrospinal fluid that surround and bathe the cerebral cortex are also present within and around these archaic brain structures. Cerebrospinal fluid surrounds the brainstem and fills the hollow of the *brainstem canal*. The surfaces of the modular components of the limbic system (the hypothalamus, amygdala, and hippocampus) interface with the *ventricles*, the interconnected system of large cavities filled with dazzlingly clear cerebrospinal fluid. There are some very intriguing channels, grooves, and hollows that connect the brainstem directly to these higher structures. Perhaps here as well the pulsations of *Light* are propelled along the surfaces of this astoundingly, blindingly, scintillatingly clear (Have I made *myself* clear?) liquid medium.

At this point we must make somewhat of a metaphysical leap—to talk about the *agency* of the *Light*, the *will* of the soul—in order to understand how the lightbody participates in the development of the neuronal structures of its own nascent brain. We know that we *have* will, not simply because it feels like we have will and appear to use it regularly—for the diehard skeptic would say that that may be an illusion, an epiphenomenon, or an elaborate ruse. If this highly complex lightbody can be proven to exist, we would be able to infer that it must have a use because no biological system can afford to invest such an inordinate amount of time and energy into the production of any such structure or phenomenon that is useless to its survival. So—is it possible to deduce a use for consciousness?

As my old friend René Descartes used to exclaim, in his cups: *"Cogito ergo sum!"* which translates: "I think; therefore, I am!" What I presume he meant to say was: I am *experiencing*, therefore I must *be* that which has the capacity to experience. The capacity to experience is the defining characteristic of consciousness; therefore, I am or have consciousness. The nature of my experience—including my conscious drives, moods, and emotions—can be shown to fluctuate with the biochemical state of my brain. Moreover, my brain has evolved complex features, such as the primary visual cortex, apparently solely dedicated to the production of my conscious experience. The neurons within the primary visual cortex are too cunningly arranged for presentation and fire far too energetically to be mistaken for non-conscious, cybernetic biocomputational circuits. Such lavish caloric expenditures would not be tolerated by evolutionary pressures if these presentation cortices did not have a clear and compelling *usefulness* to the body—in terms of its survival—commensurate with the expense. Usefulness is defined as "having a beneficial effect." Ergo: my very expensive consciousness must produce a great number of beneficial real-world effects. The lightbody has *agency*. The soul has *will*!

Consciousness, therefore, simply cannot be some useless, passive hanger-on within the brain. The fact that such a finely crafted sensorium with such intricate neuronal presentation structures exists in the first place is in itself indirect proof that the recipient of that presentation must be a valuable member of the executive team, the one responsible for directing all actions that are not accomplishable by means of unconscious processes and reflexes. In mediating with the outside world, conscious awareness can direct or avert the body's eyes and limbs toward or away from various objects of desire or aversion. Internally, it has the choice whether or not to place itself (by means of its attention) within certain cerebral regions and to detach itself from others—whether to feed certain imaginal apparitions that occur to the mind in the form of particular resonance patterns of *Light* or to dampen and repress them. There has been a great deal of speculation regarding the specific biological mechanisms that consciousness might employ in accomplishing these ends (voluntary movements of the limbs, complex coordinated actions, etc.), but a thorough review of the literature would send us very far afield. Instead, let's take a brief look at one likely candidate—the Tuszynski microtubule module—as explained by Jeffrey Satinover in his book, *The Quantum Brain*.

The question now before us is how the indwelling consciousness, in the form of the lightbody, manifests its will in bodily actions and willful shifts of attention. I have offered an argument that the will *must* be made manifest *somehow*, for without willed action the entire sensorium experience would have to be viewed as a useless and capricious waste of valuable energetic resources. I have indicated that chaos theory may be invoked to show how small, well-placed energetic inputs can result in massive shifts in the resonance patterns of nonlinear systems such as the brain and that this property of chaotic systems points to a likely mechanism by means of which the lightbody effectuates changes of mind state and initiates movement in the body.

The triggering mechanisms that allow the neurobiologically organized quantum electromagnetic field—which I call the lightbody—to influence the physical brain must by definition be sensitive to minute quantum effects; therefore, it is almost certainly the case that some form of bioenergetic-biophysical transduction structure exists at the sub-neuronal scale capable of affecting the firing patterns of individual neurons. One such theoretical mechanism is the Tuszynski microtubule module. Microtubules are extremely long, microscopically thin hollow tubes (with an interior diameter of only 14 nanometers) that serve many functions—as structural elements in all types of cells—and that are particularly prevalent in neurons. Tuszynski theorizes that the two types of dipole tubulin molecules that make up the helical lattice structure of the microtubule are capable of propagating informational waves through kinks in the "frustrated lattice structures"—and therefore "can function simultaneously as electromechanical signal transducers (longitudinally) and spin glass-type parallel computers (circumferentially)." Others suggest that it is the ordered state of the pure water molecules within the microtubules that may allow them to propagate oscillations of light or sound waves in quantum-coherent patterns. However it turns out that the *Light* of awareness is able to enact its will within the physical body through the neurobiological mechanisms at its disposal, what is abundantly clear is that it *somehow* manages to do so, and, for the reasons cited above, the microtubule seems currently to be the most plausible instrument for effecting this downwardly directed spiritual transduction. Unfortunately, the precise and definative mechanism by means of which the will is expressed must remain for now in the realm of mystery.

Willy nilly (either by an act of will or not) conscious awareness gravitates toward the focus of *attention*—and attention is the royal road to the further growth of consciousness. We learn what we pay attention to. Conscious awareness is able to withdraw itself, by means of its attention, from situations (and neuronal structures) that it finds unpleasant and focus itself upon those situations and structures that it does find pleasant—or interesting or simply novel. A newborn baby demonstrates this function of attention quite plainly, engaging enthusiastically with any new, brightly colored, or interesting object it encounters, as well as familiar objects with established positive associations. The "objects" it encounters before birth are internal—attention is focused upon fascinating new body parts, new mental capacities, sensations, and moods.

This development by exploration is of course not entirely open-ended; the genetic code dominates all aspects of mental and physical growth. Genetic mechanisms enlist the invaluable aid of conscious awareness in fine wiring these newly emerging neuronal structures by rewarding certain investigations with pleasant neuropeptide releases and discouraging others it deems are "off track" with the release of doses of equally *unpleasant* neuropeptides. The body sets up its system of rewards and punishments early on. Prenatal consciousness responds to these perceived "good and bad angels" in the same way that it will continue to do for the rest of its existence—by compliantly obeying the body's dictates or suffering the consequences.

Although there are definite genetic constraints upon the growth and development of the brain, the indwelling consciousness is a powerful force that shapes, to a large extent, its own destiny. The decisions it makes, where it places its attention, all have real and lasting effects upon its eventual form. During this period of intense isolation, each nascent being is its own mentor and guide. Giving an unexpected twist to the old "nature vs. nurture" debate, within the womb *you* are the parent to your own developing brain. You work with what you have, genetically speaking, but you work diligently and without rest to create the most attentive, dynamic, imaginative, and responsive brain that you are capable of creating. And it's all child's play.

A SUMMARY OF CHAPTER SIXTEEN

Jason Brown's theory of microgenesis is presented as a framework that successfully integrates the three disparate fields of evolutionary biology, developmental physiology, and cognitive neurobiology. His theory posits that thought moments are continuously generated from pacemaker neurons within the brainstem, traveling through sequential layers of brain structures in the same order that they were laid down in evolutionary history and in fetal development.

From an initial phase of pure wakefulness in which the self has not yet been formed, awareness travels from the brainstem into the limbic system and temporal lobes where it manifests as dream consciousness. The thought moment next experiences a division into self and world surround within the neocortex, finally emerging as a completely realized analytic perception.

One of the tacit implications of Brown's theory would seem to be that the developing fetus also depends upon the presence of an indwelling consciousness to fashion the finely adapted neuronal structures of its brain and that the genome would be incapable of orchestrating the production of such nearly infinitely complex structures in its absence. Since children and adults are clearly able to alter the course of development of their own fine brain structures through life experience, study, and skills practice, this suggests that the fetus may also be responsible for consciously developing its own emergent neuronal structures while in the womb.

The chapter concludes with a discussion of a possible psychophysical mechanism that consciousness might employ to impose its will upon the brain and the body. The susceptibility of chaotic systems such as the brain to massive, systemic reorganizations—from small, well-placed input triggers—indicates that the interface between the conscious quantum field of the mind and the stubborn physicality of the body might be found within tiny cellular structures called microtubules. These structural elements are very prevalent in neurons and may be capable of sustaining oscillations of light in quantum-coherent patterns.

CHAPTER SEVENTEEN

ONCE MORE, WITH FEELING

There has always been something disquieting about this Buddhist *Abhidharma* concept of the *thought-moment*, the term for the temporarily existing flash of consciousness that arises from the process described in chapter six—that I called *egogenesis*—which is also the philosophical framework for Jason Brown's theory of microgenesis. I had the same visceral reaction upon first reading Brown's book, *The Self-Embodying Mind*, dreading that here indeed was clinical evidence in support of this most alienating of ontologies.

That I may be deeply deluded is one thing—I can handle that. It's actually good news to find out that one's understanding of reality is somehow incorrect or incomplete. It implies that there is still the possibility that with diligent and persistent practice one will be able to break through the veils of illusion to find, on the other side, what one can only hope is a more substantial and permanent existence. But to discover that at the very core, where a solid Self should be, is nothing but a stream of wispy ephemera drawn into existence by unknowable forces—this constant coming into being and dissolving away—now that's much more difficult to take.

Something just doesn't feel right about this analysis—and it occurs to me to consider the fact that these teachings are based upon conclusions drawn by individuals while in deep states of meditative *introspection*. It may indeed be true that if you focus your attention inward, back upon the point of origin of your own consciousness, you will see there precisely what is described in the *Abhidharma*, an illusory *self* forming from flashes of awareness emanating from the plenum-void. To get at the real story, however, you can't depend upon introspection alone—just as you cannot understand the entirety of an oak tree simply by studying the acorn from which it came. If you were to look deeply into an acorn, you would find there nothing much more than a packet of DNA, indistinguishable by gross observation from that of a field mouse, a shark, or a rosebush. If you were to conclude from that observation alone that these animals and plants therefore do not exist, you would be overstating the case considerably. They do not exist in and of themselves, it's true. Their existence is subject to causes and conditions—they are temporary manifestations that emanate from a common biological process—but they are indeed entirely real.

In the contemplative traditions, practitioners give a great deal of credence to the insights gained through this type of deep introspection—particularly if, as in this case, the theoretical model has been tested repeatedly through the millennia by innumerable reliable witnesses with impeccable spiritual credentials—their findings unfailingly consistent and universal. What cannot be known by this introspective process, however, is whether this observation of the division of consciousness into successive discrete moments of incarnation reveals a basic property of reality itself, in the sense of its being a physical or metaphysical law—implying that consciousness could not be otherwise configured because the universe *itself* is somehow divided up into these discrete temporal units—or whether it simply uncovers a stubborn artifact of evolutionary biology. Could it be that when meditators retrace with their awareness back through successively deeper planes of consciousness and catch a glimpse of the origin, what they are experiencing in that moment is not the universal, cosmic source of all being, but merely the stroboscopic flashes from a rudimentary neurobiological mechanism that happened to have evolved hundreds of millions of years ago in this particular lineage of life forms—the *Light*-pump in the brainstem?

Before you decide to go into a panic over this new theoretical construct, perhaps it would be wise to do a bit of *extrospection* first. Look around. Soak it in. What does it feel like to be alive? You exist in the present—that is clear enough. The past is a fading memory, and the future a blurry projection; but is this present moment absolutely paper-thin? Does your consciousness feel like something that exists only for a split second and then is gone, replaced wholesale by something or *someone* entirely new?

I must reject that view, largely upon aesthetic grounds. It just doesn't feel right! For that matter, it doesn't hold up to logical analysis either. The visceral experience of continuity through time can be demonstrated to have as strong a neurobiological foundation as the punctuated emergence of *self* postulated by the theory of microgenesis. Continuity of existence through time is by no means antithetical to consciousness—it is simply much more difficult to create the conditions for sustaining it in a physical brain. The *Light*-pump, whose job it is to thrust pulses of awareness into sensory maps, is a relatively simple mechanism for evolutionary forces to build from scratch. It's like the Stanford Linear Accelerator, a one-shot atom smasher, easy enough to construct, relying on brute force to slam a stream of electromagnetically propelled particles into a stationary target.

To get things really moving, you need something like the Large Hadron Collider—slated to go online in May, 2008 at CERN (Conseil européen pour la recherche nucléaire)—a massive particle accelerator of a *circular* design. Much more challenging to build, much more exacting to operate, yet those extra spins around its Super Proton Synchrotron's 27-kilometer circumference superconducting dipole magnetic containment field provide an expanded timeframe to maneuver in—enough time to work those particles up into a real frenzy.

The neurobiological correlate of the synchrotron in this belabored analogy is of course the neural resonance chamber—which brings us back around, rather awkwardly, to our old standby, the *Light*-violin. Just as a violin sound box is made up of countless interconnected fibers of fine-grained wood, the neural resonance chamber is made up of billions of simple neuronal oscillators that behave in concert as a complex, but ultimately passive medium for sustaining sense data.

This allows the sense data to evolve, develop, and amplify, so as to form a more compelling presentation, by extending its activity into the dimension of perceptual *time*. In the same way that a violinist must provide ample time for a note to build into a compound wave of subtle tonality and then allow more time for the note to decay—to relinquish the vibrations of that note as the musical passage progresses—the neural resonance chamber also requires a generous time dimension to create some of its more elaborate sensory effects.

Perhaps the earliest forms of sensorium consciousness were only capable of producing "virtual Psyches"—momentarily existing conscious beings that were thrust into primitive sensory maps to make a quick and cursory reading of the environment, give a knee-jerk reaction, and then fade immediately into oblivion. There could even be some survival advantage to the body in maintaining Psyche's naïveté in certain situations. If you want to get an honest opinion out of her, it may be best just to go ahead and pop the question, right there on the spot. Often the first response is the truest. In situations of fight or flight, having an untutored Psyche enter bright-eyed and gullible onto the scene in each new moment might ensure an immediacy of reaction that any additional time for reflection would only muddle. The brain holds this atavistic trump card in reserve, occasionally authorizing the release of cascades of neuropeptides during climactic episodes of mortal combat that force consciousness back down into this more primitive but also more effective command center, the cockpit of the fighter jet itself. Relying on seat-of-the-pants intuitions and hardwired quick reflexes, fresh-faced Psyche becomes the scrappy, youthful ace pilot, more likely than the mature, contemplative commander to get us out of this dogfight alive.

Other than this, I can't imagine much of an evolutionary benefit in keeping Psyche so taut and troubled. Any conjectural advantage to such a restrictive mind state would only apply in these life-and-death situations. As long as she is not permitted to see through the illusion of the ego construct—to catch a glimpse of the huckster behind the curtain posing as The Great and Powerful Oz—it seems that under normal circumstances the body has every reason to want to extend each one of Psyche's visitations, to allow her some time to settle into her skin. Ideally, the body would be able to meld these separate iterations of Psyche into a single, collective, unified experience of *self*.

Now that feels more like the body-mind I know. Yet how could such a remarkable feat actually be accomplished? Perhaps the body allows this *community of mind* to hold its newcomer meet-'n'-greet within the cozy and commodious neural resonance chamber.

The persistence of our personality through time is the one concept we cling to most desperately, and the prospect of its loss as a consequence of amnesia or senility is a very frightening contemplation indeed. The theory of microgenesis maintains that the appearance of a persistent *self* and a *personality* is due solely to consciousness receiving consistent input time after time from stored memories, during its repeated emergences through the successive layers of brain structures. I do not dispute this as a major contributing factor, but I hold that the palpable *sensation* of the continuity of experience derives from something quite different and equally real: the thorough interpenetration of these thought-moments into a single, unified, *personal collective consciousness* within the neural resonance chamber—experienced as the *continuous personal self* or *personality*.

Whenever such a theoretical mechanism is postulated, there needs also to be offered a credible explanation for its initial entrance onto the evolutionary battlefield—very often its current adaptive advantage is not the same as that which prompted its original deployment. In this case, we would need to determine what basic, robotic functionality this transitional form of neuronal resonance first afforded the nonconscious, mechanistic brain of an early cybernetic ancestor. Certainly such an intermediary function is not difficult to envision. Even the most primitive adaptive systems of artificial intelligence—which would include the cybernetic brain systems inferred for life forms that predate sensorium consciousness—already require the integrated functioning of feedback and feedforward loops to reprocess sense data, providing the cybernetic brain with a basic form of short-term memory. This simple reentrant data processing module would be nearly identical in form, though not in function, to the hypothesized neuronal resonator.

Only with the advent of sensorium consciousness could neuronal resonance become the mechanism for more *complex* pattern recognition and classification of sense data into the categories of objects of desire and aversion. From there—gliding easily straight through a series of hypothetical intermediate forms (!)—neuronal resonance has permitted the production of progeny of more and more highly refined aesthetic sensibilities. These precocious offspring ultimately evolved into a most peculiar species of *human being*: the eminently conscious, suave and sophisticated *connoisseur* who, blindfolded, can place and date a single sip of Cabernet Sauvignon as a late harvest '83 from the Médoc region of Bordeaux—although I wouldn't necessarily trust his judgment in the cockpit of a fighter jet.

Homo aestheticus, that's our lineage—the human race of sentient beings who have taken the basic poisons of desire and aversion, which were the precious gifts of our ancestors, mixed them with a healthy dose of delusion, and now revel in an artful world of constant creation, aesthetic connection, and uproarious beauty. But beauty takes *time* to develop—it's a product of deep resonance—and resonance is a property of vibration—and vibration is something embedded in *time*.

Seventeen years ago I wrote the following three paragraphs in the introduction to an unpublished article entitled *Keeping Body and Soul Together: A Theory of Neural Resonance*, in which many of the ideas in this book were first explored. I hadn't reread these paragraphs until very recently and could only vaguely remember having written them. What was most surprising to me was—after all these years—how much I still sounded like the same person.

> *Our bodies exist in a vast sea of vibrations. From as far away as the great galaxy in the constellation Andromeda or as nearby as a whisper, the senses receive a continuous sampling of the vibrational fields that surround us. Through transformations of unimaginable complexity, these waveforms are reorganized, re-creating a living model of sight, sound, smell, taste, and touch within the mind.*
>
> *Everything in the physical universe displays some form of periodic or quasi-periodic motion: the quantum vibrational probabilities of elementary particles, the movement of energy fields, the propagation of light and sound, the rhythmic cycles of nature, and the orbits of planets, stars, and galaxies. Human beings exist on a scale roughly intermediate between the infinitesimally small and the unimaginably large. Taken in this context,*

it is almost inconceivable that the structure of the human brain—the subtlest and most intricate mechanism in all of creation that houses the conscious mind within which the world is made manifest—would not in some fundamental way reflect this universal phenomenon.

And that is precisely the thesis of this article: that there exist specific physiological mechanisms within the brain that translate the received vibrational patterns of the senses into a resonant field that is immediately accessible to the consciousness of the incarnate soul.

I certainly recognize a persistent *self* in these earlier efforts, not too dissimilar from my most current microgenetic iterations. I'm not so worried anymore by the "temporal claustrophobia" that can be brought on by *Abhidharma* study. The body has its mechanisms that once established are difficult to circumvent—coughing up these independent Psyches unceremoniously, like so many hairballs—but the soul has ways of undermining the body's primitive punctuation and has learned how to express its voluptuous fullness and continuity by other means. Evolution, like marriage, often involves the art of compromise. Cupid and Psyche have worked things out the best they can, adding new rooms as their increasing fortunes have allowed onto the original "fixer-upper" they received from his parents as a wedding gift. Cupid is a talented handyman and has made some nice improvements for his wife—in particular, expanding the reception area and brightening it up with a new skylight—knowing how she loves to entertain. They are hampered by some basic inefficiencies in the original wiring, but she is not about to let him tear up the walls hunting down the problems, so there's really nothing to be done about that.

The largest—and most neglected—of the seven books that make up the *Abhidharma Pitaka* is the *Patthana* (the *Book of Causal Relations*). Here we find an antidote to the linear analysis of consciousness so prevalent in the other books. There are twenty-four conditions listed that shape conscious experience and among them are many that bespeak a high degree of interconnectivity, back-and-forth movement, and circularity indicative of resonant, nonlinear systems. These include: *reciprocity, support, contiguity, simultaneous origination, immediacy, nutriment, association,* and *non-separation*. The main purpose of this catalogue in the *Book of Causal Relations* is to show how each thought-moment is conditioned by innumerable cofactors—how any linear analysis of the trajectory of consciousness is quickly derailed by countless force vectors arising from untold sources.

Of particular importance here are the conditioning factors of contiguity and immediacy, which refer to the way in which a thought-moment is conditioned by the thought-moment immediately preceding it or how a state of mind is conditioned by the immediately preceding state of mind. In this book we find the earliest intimations of what I would describe as a *cooperative* stream of consciousness—one where the freshness of the new is warmly embraced by the established community of mind.

If you are running a corporation, there are two ways to treat your "human resources." The first is to hire a series of temporary workers who do as they are told, are offered no benefits package, are milked of their last drops of productivity through fear and intimidation, and are summarily dismissed at the end of the day, worn out and ragged.

The other way is to develop a friendly and welcoming corporate culture, offering your employees stock options and childcare, granting the members of each department creative control over their own projects, throwing fun company picnics, and encouraging participation on casual Fridays. What is gained from this management style is a surprising increase in productivity, a growing sense of loyalty to the corporation, and a positive, gung ho attitude.

The bottom line is that both corporate strategies work well enough for their genetic shareholders—ancient alligators still successfully complete their hostile takeovers at the edges of waterholes, bringing down more progressive corporate citizens without batting an evil eye. But in the long run, if souls have any say in where they play their next gig, I imagine there may be some spirited jockeying for a position in one of the more hospitable, worker-friendly environments. The talent moves where it can do its best work—and word gets around.

In order to create this relaxed atmosphere, what is required is the invention of *time*, and once time has been invented, it needs to be expanded into a livable dimension of at least a few seconds—and if possible, more. Is that really how it works? Is time an *invention* of consciousness? What is time composed of? Is it anything real? I am of the view that time, as we experience it, is an illusion—one made possible by and only discernible through *sympathetic vibration*—an entrainment of the vibrational patterns of one *conscious* medium with the vibrational patterns of other (presumably *nonconscious*) media that surround and interpenetrate it. This fundamental illusion of time is the framework that allows for the establishment of all of the other illusions of representational reality depicted within the sensorium. There is no cause for alarm; *illusion* here is simply a "term of art." I am by no means implying that the universe is not actually there or is in some sense fundamentally different from how it appears to us to be. But there is most definitely an illusion taking place, and that can be proven, not beyond, of course, but within the shadow of a doubt.

We exist in the present moment—but that moment is not absolute or still; it has momentum—it is moving through time. Or does time only *appear* to move because *things* move within this single, unique, and ubiquitous moment, and the changes in spatial relationships that result from this movement *imply* the existence of a separate dimension of time? Change happens, and different degrees of change take corresponding amounts of time to happen—the implication being that there must have existed separate times prior to this one to pair with each of the self-evidently different previous states. Since there is no room in the now for all of these states, we infer a temporal dimension, a past into which moments of present are shunted once they have appeared in the now. Therefore, we naturally surmise, the universe must consist of a very narrow present and a much more ample past. We find evidence of there having been other moments by referring to archives—those held in memory, in myth, in keepsakes of bygone days, in dusty diaries and photograph albums. From these observations we jump to the seemingly logical conclusion that there must in fact be an infinite number of times, all laid out in sequence, contiguous like space but unidirectional—always going from the past through the present toward the future. But if these separate moments that constitute the dimension of time actually existed, where are they now?

Time is a paradox. It is what allows for—but is also somehow created by and depends for its very existence upon—the oscillations of matter and energy. Without a tick and a tock there could be no clock. If you can imagine making the thinnest possible slice through the presumptive four-dimensional space-time continuum, so that the three-dimensional world was cryogenically frozen in an absolute stillness between past and future, you would see and hear nothing whatsoever. Colors could not be distinguished, for light can only reveal its particular wavelength if it is given enough time to complete at least one repetition of its cycle of propagation through the electromagnetic field. Musical pitch would be impossible to determine, even the presence of sound itself, for sound is due to the repetition of variations in air pressure propagated in waves through space and time. And there's the rub. What constitutes the *repetition* of anything?

You can sit on the shore and count waves, to test the folk theory that the seventh wave is always larger than the previous six. And if you are not distracted by the flock of seven pelicans gliding over the water, you will have no difficulty counting the waves. You will hear the crash of each one breaking, you will feel the warm water tickling your toes, as each succeeding wave fans out over the smooth white sand, pauses briefly, and then retreats back into the ocean. Even a child can count these waves easily enough, but what is it that allows us to jump to the conclusion that any one of these "wavelike" events has anything at all to do with the previous one? All seven waves were different; no one was completely comparable to any other. They were shaped somewhat differently; one may have been larger or smaller than the next. The interval of time between their crestings may have varied from set to set.

The seven pelicans are fundamentally different from the seven waves, but there is something deep inside our brains that has no difficulty making the translation. Counting separate objects in space is not the same as counting the risings and fallings of a single medium through time. When you are done counting seven pelicans you still have seven pelicans left. When you are done counting seven crashing waves you only have the one last wave left. There never was any more than one wave crashing at a time. Oh, but the brain can't be bothered with these philosophical musings now. It's too busy counting. The brain is incessantly counting and displaying the results of its counting to consciousness. Analyzing frequencies of sensory input, comparing and contrasting whole number ratios with a variety of neuronal oscillators, performing harmonic analyses of complex waveforms in neuronal resonators—doing this type of "musical mathematics" in all six sense modalities is the full-time job of the neural resonance chamber.

In just this way, the body-mind sits upon the shore and counts the world into existence. The world gives off waves of vibrations of all frequencies in a variety of "artistic media." Modern sense organs evolved originally from archaic body parts fortuitously sensitive to particular modes of vibration and from which the organism was able to glean pertinent survival information. Now equipped with a full complement of highly developed sense organs and sensory cortices, present-day animals are capable of receiving, resonating, analyzing and integrating a large number of continuous streams of complex vibrational data. Employing vast armies of "tabulators," the sense organs count, in various ways, the number of vibrations per unit time, eventually assigning standardized *qualia* (such as "trumpet B-flat," "midnight blue," or "lavender scented," depending upon the medium) for certain numerical categories of vibrational sense data. The counting is most often done indirectly—the brain and body having evolved many algorithms, tricks, and shortcuts to make the counting easier.

The primary sense organs are arrays of "tuning forks" of one kind or another, each responding sympathetically to a particular *frequency* of environmental vibration. A frequency is by definition an accurate *counting* of the number of waves cresting per second. Only hair follicles of a certain length and at a particular location in the conical, fluid-filled spiral chamber of the inner ear, for example, will vibrate sympathetically in response to a given *frequency* of sound wave; the signals from cells at their bases are transferred to the auditory cortex where they are experienced in the brain as a corresponding musical tone. Three types of photopigment molecules within the cones of the eye respond preferentially to one of three *frequencies* of light. The relative signal strengths from these three types of cones are analyzed and displayed as a corresponding color in the visual cortex. Olfactory receptors in the nasal passages—operating as tiny molecular spectroscopes—respond only to molecules whose primary vibrational modes are of a particular *frequency*, thus assigning to each molecular substance its corresponding smell, experienced in the rhinencephalon (the "smell-brain") within the limbic system. This phylogenetically older mammalian brain center is the seat of emotion and intense sense memories. This explains why odors and perfumes—which are actually the vibrations of a sampling of the very molecules that exude from our bodies and our environment—can put us in such direct contact with all of our ancient animal passions.

A fascinating account of Luca Turin's discovery of the vibrational basis of olfaction can be read in Chandler Burr's wonderful book, *The Emperor of Scent: a Story of Perfume, Obsession, and the Last Mystery of the Senses*. For me it was a revelation—and a confirmation that in fact *every* sense modality is indeed a manifestation of some aspect of the vibrating universe, perceived and translated by other aspects of that self-same universe into a resonant field of electromagnetism within the brain and body, *Light*-ing up the multisensory theatrical display of the sensorium. The sense of smell must have been the earliest of the senses to have evolved. It occurs through the direct and intimate contact of the flesh of a living organism with the substance of its environment. The ability to distinguish the smell of food from poison, mates from predators, the familiar home from the rival's territory—scent is at the very root of sentience.

Beneath even this level of basic sensation, there are strong indications that resonance itself may turn out to be the most fundamental property of the universe—according to proponents of String and Superstring Theories. Elementary particles that make up the quantum fields that make up the atoms that make up the molecules that make up the *ad infinitum* are now to be viewed as musical notes or "excitation modes" of even more elementary *strings*. Like violin strings stretched under tension, which when plucked produce particular notes, these infinitesimal strings of space-time vibrate the physical universe into being. These are strange strings indeed—absolutely one-dimensional, yet formed into open or closed loops that are tiny beyond imagination. A violin string is approximately ten-decillion (10^{34}) times the length of the average elementary string. In a most ambitious gesture, superstring theorists propose to link each *boson* (a force-carrying particle) to a particular *fermion* (a particle that propagates waves of matter), thus describing a magnificent, mathematically perfect, vibrating universe of *supersymmetry*. Even the Hindu Vedas (written circa 1500-1200 B.C.E.) concur with the upshot of this analysis, having asserted all along that the universe is in actuality the all-encompassing vibrational sound of the *Aum*, which was with Brahma and inseparable from him since the beginning of time.

There is a well-known aural illusion called the "missing fundamental" that gives some indication how at home the brain appears to be in this universe of resonance. A complex tone, such as that produced by a violin, comprises a full series of a dozen or more prominent harmonic overtones, all integer multiples of the fundamental frequency, each apparently processed independently by the brain but combined in perception into a single fundamental tone—the harmonics simply creating what we call the instrument's *timbre*. A striking illusion occurs when two or more sine wave oscillators connected to a speaker are made to generate pure tones from a given harmonic series. The brain correctly reduces the perception to a single sound at the fundamental frequency even though that frequency is not itself present.

It was once believed that the "missing fundamental" must have been physically reintroduced into the waveform by the force of the combined harmonics "driving" the fundamental tone as a standing wave distortion within the cochlear fluid of the inner ear, but this has been convincingly demonstrated not to be the case. Another notion, that the fundamental is "calculated" by the brain through a computational frequency analysis of these harmonics and then reintroduced as a new perception, seems unlikely, given that "missing fundamentals" are not naturally occurring phenomena. The fundamental tone is always present in the sounds of nature and is normally of higher amplitude than any of the upper harmonics. It is doubtful that such a complex skill set would have evolved within the human brain during the last half-century simply to cope with the annoying music from those cheap transistor radios introduced in the 1950s—I cannot think of a strong evolutionary advantage to our being able to discern the sounds of a bassoon emanating from their tinny, miniature speakers.

If one is unable to locate a mechanical "fundamental wave driver" within the cochlea—whose trumpet shape is certainly the most obvious analogue to a musical instrument already known to be capable of producing such effects—perhaps the "missing fundamental" is actually produced mechanically (or *neuromechanically*), not within the cochlea but within the auditory cortex itself. The theory of neural resonance maintains that in order for the auditory cortex to analyze and *present* a quantum field *display* of complex rhythms, harmonics, and musical chords, it must itself be arranged in a manner precisely analogous to a musical instrument's sound box—being comprised of millions of physically interconnected neuronal resonators arranged in orderly ranks. Therefore, the combined harmonic wave driving may actually reintroduce the fundamental tone directly into the cortical oscillations through a standing wave distortion similar to that hypothesized for the cochlea—which would mean that there is actually no "illusion" at all! The "missing" fundamental tone would have been reintroduced as a new oscillation right there in the auditory cortex.

I have mentioned before how in scientific research early choices in vocabulary and conceptual frames seem to constrain the imaginations of later investigators. One of the most persistent examples of this is the equation of the neuron with the computer circuit, which leads to the assumption that the brain must fundamentally be a data processor, performing some sort of hypercomplex Boolean algebra with these billions of integrated neuronal circuits, AND, OR, and NOT logic gates, switches, and relays. Neurons are interconnected, they pass signals to one another, and at any given time they are either firing or they are not firing, this "decision" governed by rules of cell physiology. Computer circuits are also interconnected, they pass signals to one another, and at any given time they are either on or off, that "decision" also governed by the rules of their operating system. Neurons are *ipso facto* biomechanical computer circuits. Well...

The trouble with this train of logic is that it *starts* with the *conclusion* that the brain is ultimately binary in nature—that what a neuron is doing is deciding, based upon its input, whether or not to send a "yes" signal on to the next neuron down the line. The absence of a "yes" signal means "no." These two possible states can be symbolized by a one and a zero to make the computer analogy complete. From just such a basic bootstrapping algorithm, the most complex computer code can be written and stored, in a data file consisting of a single, extraordinarily long sequence of ones and zeros.

The whole computer paradigm is *so-o-o* 20th century. Rather than conceptualizing the cerebral cortex as a vast assemblage of individual neurons, each one behaving like a rudimentary, isolated, binary on/off switch, imagine instead richly integrated clusters of neurons expressly designed to *reverberate* their signal impulses—simple neuronal oscillators rationally organized like ranks of organ pipes into more complex and elaborate neuronal resonance structures. It is the *circle* of neurons, rather than the *single* neuron, that should be viewed as the basic unit of organization for the neural resonance chamber.

If this is indeed the case, then the neuron is not analogous to a computer circuit—it's part of a *circus*; and not just a three ring circus, but a 300 billion ring circus. It's the Greatest Show on Earth, where lions and tigers and elephants are made to appear and disappear, acrobats perform their synchronized routines high up on the flying trapeze, raucous caliopy music fills the tent, and the audience sits in rapt attention. Binary computers are excellent devices for manipulating, storing, and retrieving vast amounts of data, but unless they are able to *display* their findings in such a way as to grab the spectator from his fit of abstraction, their complex analyses are all for naught.

The computer paradigm rests on the assumption that neurons are processing pure *information* and consequently provides no clue as to how these bits and bytes and gigabytes of raw sense data might be transformed into anything even remotely resembling the rich and vivid *experience* of sensorium consciousness. Neuronal resonance is for this reason alone a much better model: it is much better at *modeling*. It provides the brain with the means for molding shapes and forms in three dimensions from the plastic material of *structured electromagnetism*. Together, the brain and the mind can manipulate these electromagnetic forms, meld them, compare them with other forms, associate and disassociate them, classify and reclassify them—and finally solidify them by baking them in the oven of long-term memory so as to be able to retrieve them later at will.

In order to have a completely open exchange between the three functions of sensation, memory, and analytical thought, it would be very helpful if they all happened to be using the same operating system. *Neuronal resonance* may be the activity of the brain that ties all three of these aspects of mind together. In earlier chapters I have mentioned the Buddhist conception that thought, memory, and other forms of mental visualization operate through a sixth sense organ (also referred to as "mind") and that this conception forms the basis of their view of the equivalence and parity of all sense perceptions and mental formations—an analysis pursued principally to show that they are *all* equally *illusory*. In evidence of this, we have already seen how the frontal cortex, associated with rational thought in humans, is organized with the same laminations and cortical columns that one finds in all other sensory cortices.

The brain sustains our memories and, in order to do so, is continuously reconfiguring itself into the ultimate *mnemonic device*—which stands to reason, since a mnemonic device is something that aids in *memory*. Giordano Bruno, a Hermetic heretic of the Italian Renaissance, is *remembered* for his writings on the art of recollection, in which he described creating in his mind an elaborate mnemonic device in the form of an infinitely expandable *Palace of Memory*. Any fact or concept that he wished to remember could be encoded into an appropriate object form, concretely visualized, and placed in his mind in a special location, surrounded by associated symbolic concepts, in a particular room in one of the many wings of the palace.

Whether we intend to build such palaces of memory or not, the brain seems predisposed to arrange its collection of recollections in a similar way, leaving us to hunt about the palace grounds in vast impressionistic landscapes of associated memories or to stumble through dimly lit hallways of the distant past to retrieve a variety of treasured objects of mind. Regardless of whether we take the time to catalogue our own memories properly by careful reflection in the light of day or allow them to sort themselves out each night in dreams, the brain appears to have thought up this system of memory by association all by itself, long before Giordano Bruno described his own artful refinement.

Perhaps the most important similarity between Bruno's artificial mnemonic device and the brain's functional analogue is that both systems depend upon objects having discernible shapes and recognizable features. I have used such "palaces of memory" myself in the past, to cram for exams in college. I also use a similar system to remember people's telephone numbers by transposing the digits into particular letters of the alphabet and then into words and then into *objects* that I arrange in my mind in an odd and therefore memorable juxtaposition. After forming an image of the person in question surrounded by these peculiar objects, I find I am able to retrieve the image easily from memory and decode the person's telephone number from that. What I have noticed is that my memory always begins with a vague sense of the general *shape* of the objects, and only as the memory develops and more details are recalled am I able to identify the objects as particular *things*—with names and spellings of those names—which I can then translate back into the original telephone number.

I use the term *memory mandala* to refer to the *"chaotic system"* the brain seems predisposed to use in "organizing" its memories. Traditionally, mandalas are microcosmic diagrams, such as the ones used by Tibetan Buddhists for meditative contemplation, as mnemonic devices and as teaching aids. These incredibly detailed, brightly colored paintings, called *thangkas*, contain a universe of esoteric knowledge. Deities are portrayed with particular sacred objects in each of their many hands; the objects function to remind a practitioner of the spiritual qualities associated with each deity. The overall layout of the painting is also important; it clarifies the relationships between the deities and the other beings in their separate realms of existence. The specific colors used to portray the deities and the objects also have spiritual significance.

The memory mandala is the brain's multidimensional, multisensory storage and retrieval center. It is multidimensional in the sense that memories appear to be connected and therefore retrievable by tracing pathways through a series of often impossibly contorted and contradictory three-dimensional overlays. One can zoom in upon a concept or a past experience and from there branch out in a number of different directions, traveling easily from one sense modality to another. You catch a whiff of rose perfume; the word reminds you of the name of a restaurant—*The Irish Rose*—and you immediately recall a snippet of conversation you had there last week; the memory of the face of your dinner companion calls to mind the face of a friend from college and the road trip you took together one summer; and suddenly you are off and running, lost in a rich reverie of loosely associated sense memories.

All memories, in a very real sense, are sense memories, because in order to recollect anything the mind must rummage through categories of sensuous and mental associations, often leaping from one to another of the *six* sense realms. The brain has evolved its own quirky systems for cross-indexing its memories—in often *illogical* categories. One of the main goals of our educational system seems to be to conform the wild and untamed mind of the child to society's rectilinear framework of logical categories and hierarchies (that not surprisingly ranks the rational *mind* in the very highest echelon). These categories *do* help us better organize our memories—whether chronologically, geographically, numerically or linguistically. The operative assumption is that with a brain so arranged *we* might also, like the happily humming computer to which we all secretly aspire, more easily store and retrieve life's pertinent information.

Emotions and moods determine which facets of the memory mandala are accessible at any given moment. In a foul mood it seems only foul memories can resonate. In fact, we have become very skilled at tagging our emotional baggage, so that every precious piece of it is present and visible for our consideration when we wish to wallow or revel in a particular mood. One well-known Tibetan mandala, the Wheel of Becoming, charts the "six realms of existence" that can be understood to represent the various experiential "worlds" created by our clinging to such emotional states and habits of mind.

Neuronal resonance provides us with a powerful conceptual framework. It binds together functions of mind once thought to be largely unrelated—including sensation, memory, and analytical thought. The bewildering snarl of cerebral cortex neurons is conceptually disentangled into tidy skeins, ultimately comprised of the most basic of mechanisms—simple neuronal oscillators and resonators. With such direct and easily comprehensible parallels to the resonance chambers of musical instruments, neuronal resonance offers the most visceral and accessible model for the evolution of consciousness. This more than metaphorical analogy between the brain and the violin underscores the importance of aesthetics in the processes of mind—and points out clearly that every animal with sensorium consciousness knows exactly what it likes and what it dislikes, demonstrating its preferences by the directed movement of its body toward or away from its object of pleasure or displeasure.

Neuronal resonance complements the study of *psychoneuroimmunoendocrinology* (a most ambitious term that attempts to encapsulate this truly boundless, borderless field). It provides at least *one* crucial insight: that a fundamental purpose for the "molecules of emotion" is to create within a single brain a set of independently configured *brain states*, resonance patterns functionally adapted to the specific exigencies of stereotypical moments—a brain of fear, a brain of desire, a brain of anger, a brain of meditative calm. Each particular neurochemical milieu calls up its own variation of the neural resonance chamber—in the *service* of which associated memories are brought to the fore, apposite cortical structures are engaged, and the entire system is routed toward a specialized state adapted to elicit a certain range of appropriate behavioral responses.

Neuronal resonance is what sustains all of the intricate patterns of presentation within the sensorium. It's the lazy mathematician whose abacus only glows in perfect ratios. Rings of neurons oscillating in harmonic ratios of 3:2 or 7:5 (a perfect fifth or a septimal tritone) bring harmony, tension, consonance and dissonance into the *Light* of sound, into the *Light* of light—filling the sensorium with an infinite variety of expressions and contortions of beauty.

Neuronal resonance provides the means to welcome chaos into the well-ordered brain. Chaos is a daunting word that may evoke images of madness and an end to the reign of logic, but chaos in the correct dosage is unquestionably the mind's most soothing tonic. Without a healthy smattering of nonlinear, rococo chaos softening the edges of a phase-locked 40-Hz gamma-synchronous brain (often cited in theoretical papers (Crick and Koch, 1991) as a possible correlate of attention or "carrier wave" of consciousness), we might all be mired in our endless oscillations—interminable echoes of experience rebounding through sympathetic resonance with ever-gathering standing waves—the kind of neuronal hypersynchronization that characterizes an epileptic seizure. As the Buddha advised his disciple Sona regarding the practice of meditation: "The string that produces a pleasant and harmonious sound is the string that is not too tight and not too loose." Chaos is what allows patterns within patterns to emerge—infinite regresses dissolving into white noise, from which can uprise something spectacular and new. Chaos hints at the ways in which the *will* can enter this marvelous display. A balanced "chaotic brain" is sensitive to the slightest touch of a spur, and a well-placed whisper can move the great stallion of the resonant brain forward or backward, can veer it toward this and away from that.

Neuronal resonance is how consciousness expands into the fourth dimension of time. It holds an impression of the world-surround long enough for us to compare what *is* with what just *was* before. It is the persistent evidence of a just-spent past, maintaining the steady state into which moments of newness plunge, splash, and reconvene. Neuronal resonance gathers up streams of data from miscellaneous sensory sources, forms them into leisurely moments, packs them in boxes, and shuttles them into the past. In moments of quiet reflection neuronal resonance is the function of mind that can open the gates of eternity. The *Now* becomes, like the substantive space dimensions, as palpable as North, South, East, and West. It is the wide expanse of the Center, the *ample moment*, a welcome relaxation in the tension of the bowstring that releases the arrow of time.

Neuronal resonance is a play of *Light*—wherein a self-sustaining *self* sustains the illusion of separation from the very object that it most intimately *is*. Surrounded by the overwhelming sensorium, which in actuality is a projection of its own *Light*'s body, the *self* retreats into what appears to be a hole in the middle. Mesmerized by the scintillating overtones in the upper harmonics, consciousness misconstrues the *self* to be the "missing fundamental"—the illusory bass tone it thinks it hears and believes upholds the fleeting world—the tortoise beneath the tortoise beneath the tortoise upon the back of which the universe rides.

ONCE MORE, WITH FEELING

A SUMMARY OF CHAPTER SEVENTEEN

The theory of neurobioluminescence in the evolution of sensorium consciousness proposes two new functions for the neurons of the cerebral cortex. The first function—neurobioluminescence—connects the physical brain to the quantum field of mind. The second—neural resonance—connects both the brain and the mind to the sensory world.

Each of the five sense organs evolved from archaic body structures that were fortuitously tuned to meaningful frequencies of vibration present in the surrounding media. Structures within nerve cells of the nasal cavity are sensitive to specific vibrational modes of odiferous molecules. Proteins within color sensitive nerve cells of the retina deform in the presence of particular wavelengths of light, thus signaling the cells to fire. Tiny hairs associated with the nerve cells of the cochlea resonate sympathetically with specific frequencies of sound. Our whole body responds to the rhythmic patterns of the drum, enticing us to join in the dance.

Neural resonance is what allows for the rational analysis of sense data. Thoroughly enmeshed in the mathematics of music, the brain is richly endowed with innate aesthetic sensibilities. Rational chordal structures, solid harmonics, and steady, complex rhythms proliferate here. Irrational, random inputs will find no sympathetic audience.

Neural resonance is what gives us our sense of reality, of participating fully in this vibrant, vibrating world. Neural resonance permits what would otherwise be a discontiguous tumble of microgenetic thought moments to meld into a coherent experience of a persistent, "solid" and substantial self.

CHAPTER EIGHTEEN

HIDDEN TREASURE

I'm not sure if someone can be said to be a *firm* believer in serendipity—it seems somewhat of an oxymoron—but I have the distinct feeling that fortuitous chance has played a great role in my life, most notably in my spiritual education. I do throw the *I-Ching* coins—more than occasionally—to consult the Oracle regarding crucial and often *not* so crucial life decisions. I don't know if there's anything to it, really; reading the hexagrams and commentaries certainly seems to put me in a more open and receptive frame of mind.

It's my customary habit, when I browse through a used book store, to open a book somewhere in the middle, dowsing with my thumbs, and to read from there just a few pages, paragraphs, or lines. As soon as I find something really intriguing, my first impulse is to slam the book shut, place it back on the shelf, and hightail it out of the bookstore—to stroll around the neighborhood and think about what I've just read—to watch the new thoughts jostle and parry with the stock characters of my usual ruminations.

But one day a book fell into my hands from a high shelf in the used section of a local bookstore that marked a true moment of cosmic providence for me. Although I flipped the book open in the usual manner, I found that I simply couldn't put it down, even after I had finished it. I had to go back to the beginning to read it all over again. It was an amazing discovery—like finding the Buddhist Rosetta Stone. Suddenly it all made sense!

The book was David Brazier's *The Feeling Buddha*, and although it's a marvelous read through and through, the real treasure for me was tucked away in the back of the book: Brazier's retranslation of the Buddha's First Sermon—in which the newly Awakened One delivers the pivotal, foundational teachings of the Four Noble Truths and the Eightfold Path. I believe David Brazier's work to be a crucial recovery of the original intent and spirit of these core Buddhist teachings. This First Sermon, also called the Sutra of the Turning of the Wheel of the Dharma, is the wellspring and touchtone of every Buddhist meditation technique—the root of all further flowerings of Buddhist ethics, spirituality, psychology, and philosophy. Moreover, this new translation of the sutra provides the perfect philosophical framework for understanding the spiritual implications of the theory of neurobioluminescence in the evolution of sensorium consciousness.

According to Tibetan tradition, the great spiritual master, Padmasambhava, is said to have hidden hundreds of scriptures, religious images and ritual articles throughout Tibet, to be discovered in the ripening of time by subsequent generations of masters. These *terma*, or "Revealed Treasures," are known today as the Nyingma lineage.

Could it be that a hidden treasure of another kind lies right before our eyes in the Sutra of the Turning of the Wheel of the Dharma? That is in fact the audacious claim of David Brazier, a British psychotherapist and spiritual teacher to the Order of Amida Buddha, a community dedicated to socially engaged Buddhism, who presents a convincing argument for reexamining this sutra, the First Sermon of the Buddha, delivered shortly after his enlightenment.

This is one of the most frequently chanted sutras in the Buddhist canon, and it has doubtless been so from the day it was first taught. Since its initial redaction into classical Pali, this sutra has remained for more than 2000 years fixed and constant. What has happened is that the *world* has shifted around the sutra. The vernacular language has changed. The significance of a few key terms has been altered by the passage of time. The thread of continuity in the teaching lineage may have faltered. And now, without fanfare, David Brazier quietly readjusts this cornerstone of the Buddhist edifice, revealing its original face.

It seems preposterous. But consider the ceiling of the Sistine Chapel, once hidden beneath centuries of soot from the smoke of devotional candles. Who could have imagined that taking the restorer's brush to that ceiling would reveal the true genius of Michelangelo in such a miraculous frenzy of delicate pastels? Unlike the Tibetan *terma* or the ceiling of the Sistine Chapel, the Sutra of the Turning of the Wheel of the Dharma is a treasure hidden in plain sight.

David Brazier's provocative thesis is that a radical shift occurred at an early date in Buddhist history in the definition of one key term, *samudaya*, with a consequent change in the meaning of two others, *dukkha* and *nirodha*, obscuring the message of the Buddha in this foundational text, with profound consequences to the subsequent historical development of Buddhist philosophy. He does not linger over a detailed analysis of how such an important misunderstanding could have occurred, nor does he attempt to build a completely airtight case for the correctness of his interpretation. He is, quite rightly, more concerned with showing how the reframing of this core Buddhist teaching can help individuals cope with the afflictions of life, so that they may see that their own struggles with desire and aversion are part of a natural process, and that the Buddha's gift to the world was his penetrating insight into the mechanism underlying this process and his discovery of a practice that holds the promise of true liberation.

I cannot recommend this book highly enough. Although it challenges a central dogma of orthodox Buddhism, it succeeds in realigning this sutra with the original spirit of the Buddha's teaching—the joyful, open heart of wisdom and compassion.

My purpose here is to go somewhat further than David Brazier does by tackling the fourth noble truth, which in his analysis was left relatively unscathed. I hope to show that by reframing this fourth element, *marga*, or "path," a more coherent and multilayered teaching will be revealed, closer to the one originally constructed with such infinite care by the Buddha. It is a teaching that details not only a full cosmology of sentient existence, a guide for engaged social action, and a curriculum of study for one's personal practice, but also a specific meditation technique for facing with serenity and nobility all of the afflictions of life.

I also hope to shed some light on the question of how such a misinterpretation could have crept into the orthodox rendering of this sutra. Most likely this occurred as a conflation of this text with the *paticca-samuppada* (dependent origination) material to which it bears a superficial resemblance in specific vocabulary, but to which it is not directly related in either meaning or intent.

Finally, I will offer additional support for David Brazier's new interpretation by pointing out its precise parallelism with another source—one that has been generally regarded by academics as simply a fanciful tale—the legend of Gautama's struggle with the demon Mara on the eve of his awakening to Buddhahood.

THE FOUR NOBLE TRUTHS

Vast libraries of commentary have been written through the centuries, elucidating various nuanced positions on the meaning, intent, and implications of the teaching of the Four Noble Truths. It is not really possible, therefore, to cite one definitive orthodox interpretation. When I refer in this chapter to "the orthodox view" I am describing a composite picture whose elements are drawn from a variety of sources, representing the confluence of orthodox literary, scholarly, technical and popular understandings of the doctrine of the Four Noble Truths. Nevertheless, this range of views is quite small when compared with the highly *unorthodox* translation offered by David Brazier. The following thumbnail sketch gives some idea of the degree of disparity between them.

This would be an orthodox rendering of the Four Noble Truths:

The First Noble Truth—Suffering

The Second Noble Truth—The Cause of Suffering

The Third Noble Truth—The Cessation of Suffering

The Fourth Noble Truth—The Path

This is David Brazier's rendering of the Four Noble Truths:

The First Noble Truth—Afflictions

The Second Noble Truth—The Response to Afflictions

The Third Noble Truth—The Containment of the Response

The Fourth Noble Truth—The Path

In brief, the controversy centers upon two variant and incompatible translations of *samudaya* (the second noble truth) and the two consequent views of the functional relationship between *samudaya* and the first noble truth, *dukkha*. Bear in mind that what is at issue is not how the word is translated into English, specifically; the dichotomy would be the same in all modern language translations of the sutra. In the orthodox version, *samudaya* is translated as "cause" and *dukkha* as "suffering." The agglutinative combination form *dukkha-samudaya* therefore is taken to mean "the cause of suffering." In David Brazier's version *samudaya* is not the cause *of* but rather the response *to dukkha*, here translated as "afflictions." The combination form *dukkha-samudaya* is rendered "response to afflictions," essentially reversing the traditional causal relationship between the first two noble truths. Each translation leads to a divergent reading of the third and fourth noble truths, as we shall see.

> *The noble truth of dukkha is this: birth, old age, sickness, death, grief, lamentation, pain, depression, and agitation are dukkha. Dukkha is being associated with what you do not like, being separated from what you do like, and not being able to get what you want. In short, the five aggregates of grasping are dukkha.*

With some range of variation between sects, the orthodox interpretation regards *dukkha* as an amalgam, signifying "the whole mass of suffering," "life is suffering," "suffering pervades all of existence," or some such equally dreary proclamation. The Buddha is understood to have come to a full realization of the universality of suffering (the first noble truth). He is thought to have discovered suffering's root cause in "craving"—a specific category of *desire* that is accompanied by an exaggerated assessment of the desirability of the desired object. The ultimate cause of suffering is said to be either the "craving for rebirth" or the elemental, unquenchable "thirst for sense pleasure" (the second noble truth). His profound insight is said to be his discovery that the whole mass of suffering could be brought to an end (including, unnervingly, *birth*, which is listed as the first in the series of sufferings) by the complete extinction of this craving (the third noble truth).

The fourth noble truth seems, in this version, oddly disconnected from the other three. The Eightfold Path appears here as a simple listing, without clarification, and it must be *inferred* that these eight areas of spiritual concern are to be cultivated and brought to perfection, that this would presumably lead the practitioner to a state of such purity (or disdain) that he would renounce desire completely and upon his death (never to be born again) enter final *parinirvana*.

I would ask you, at this point, to read the text of David Brazier's translation of the Sutra of the Turning of the Wheel of the Dharma, which appears at the end of this chapter, so that you yourself can weigh the relative merits of these two interpretations. And why make such a big ruckus over all this? Well, the short answer is that I think it is important that one is able to *believe* in the central dogma of one's religion. That seems reasonable enough. But I was never able to dispel my doubts about the teachings contained in this pivotal sutra until I read David Brazier's book.

Consider this line that appears toward the end of the sutra and that is not in contention in these two versions:

> *When the Victorious One had said this, the five monks were filled with joy.*

Not only the monks but the spirits of the earth and the celestial beings all *rejoiced* in the Buddha's teaching. I never experienced anything resembling joy in reading the orthodox version of this sutra. On the contrary, it filled me with dread. It presents an exceedingly dour view of life, with complete extinction as the only way out of the cycle of endless suffering. I did experience *great* joy, however, upon discovering this new, life-affirming translation.

Now what is it that makes David Brazier's rendition of this sutra so much more joyful? Here, the Buddha's enlightenment was an awakening to the best news of all. He discovered that although life presents us with many obstacles, problems, and difficulties, and although we seem enslaved by our conditioned habitual reactivity—to the craving for relief that has been prompted by these afflictions—we have the capacity to find true happiness by letting go of our attachment to the objects of craving, whatever they may be, and settling back into the peace of the present moment.

What follows now is largely a compilation of my own thoughts and speculations—informed, of course, by David Brazier's work. Many of these ideas come directly from the sutra itself, or from other Buddhist sources, as I have sought corroboration for this radical view.

DUKKHA-SAMUDAYA

Perhaps the best way to get at the distinction between the two contending definitions of *dukkha* is to understand that in the orthodox definition, "suffering," the list of items, "birth, old age, sickness, death, grief, lamentation, pain, depression, etc.," is taken to imply "the whole mass of suffering." In the alternative translation, "afflictions," each of these words, *individually*, connotes a specific instance of a particular negatively charged occurrence, which acts as a stimulus to the body-mind, provoking a response in the form of either a reflex action, a plan, a scheme, a temptation or a desire intended to subdue, distract, escape or otherwise address that particular *dukkha*.

What the Buddha was able to generalize in this most astoundingly prescient insight—into what is essentially the foundational mechanism that drives the process of evolution, the irreducible conundrum that presents the greatest challenge to our psychological health, and the impetus for the development of all forms of Buddhist practice—is that all sentient beings, including human beings, are burdened by their reactivity to negative events and that if that reactivity can be contained through the steady practice of mindful awareness, one's true nature, in the form of abiding freedom and happiness, will be revealed. Until awakening to the Dharma, humans are enslaved by the power of their own minds. When confronted with an obstacle, a challenge, or an affliction, the human mind naturally seeks some mode of escape. This is *samudaya,* which is defined as:

> ... *thirst for self re-creation which is associated with greed. It lights upon whatever pleasures are to be found here and there. It is thirst for sense pleasure, for being and not being.*

The definitions of *dukkha* and *samudaya* are intended to show the ubiquity of such afflictions and the broad range of possible responses. The first four items in the list of *dukkha,* ("birth, old age, sickness, death") are compulsory—inevitable stages in the natural course of life. They represent the cosmic scale, over which we can never have control. The next two ("grief, lamentation") denote loss of all kinds, small and large. Next come the three broad categories of negative emotions and bodily sensations ("pain, depression, and agitation"). Then follow the various frustrations of our desires ("being associated with what you do not like, being separated from what you do like, and not being able to get what you want"). Finally, the Buddha indicates that the presence of *dukkha* is also woven into the very fabric of our being, in the five aggregates of grasping: form, feeling, perception, impulse, and consciousness.

The word used to encapsulate the meaning of the second noble truth is *samudaya.* It has two roots, *-udaya,* meaning "to go up, to arise" and *sam,* meaning "with" or "together." Combining these roots produces "coming up along with" or "co-arising with" *dukkha*—hence "response to afflictions." In a carefully constructed philosophical treatise whose very subject is *causality,* Occam's razor would counsel us to presume that the numerical *ordering* of the Four Noble Truths would itself reflect this causal sequence. (A small boat *rises along with* the tide; the boat's upward movement does not *cause* the ocean to swell!) An alternative explanation has the Buddha in the role of a physician and the ordering of the truths being a description of the course of treatment: disease, diagnosis, prognosis, and prescription—although this motif is never alluded to in the text of the sutra. The explicit image throughout is of the Buddha as a victorious spiritual warrior: "the One Who Enjoys the Spoils of Victory" or "the Victorious One."

How *samudaya* came to be understood as "the cause of suffering" is somewhat baffling until it is placed in the context of another group of teachings from the vast Buddhist canon, the *paticca-samuppada* (dependent origination) material.

Also known as the *bhavachakra* (the Wheel of Becoming) or the Twelvefold Chain of Causation, these teachings offer the Buddha's analysis of the complex sequence of events that forms the cycle of rebirth or *reincarnation*. Dependent origination, in its own right, is a brilliant teaching on the machinery that binds all physical and mental manifestations, capturing us in perpetual cycles of ignorance, craving, clinging, birth, and death. But is the cycle to be brought to an absolute and final end by abandoning this world through the complete eradication of *desire*, or is this experience of incarnation to be transformed into a new kind of life, a noble, purposeful life, through liberation from the shackles of *ignorance*?

In the Sutra of the Turning of the Wheel of the Dharma, the twelve "turnings" are three sets of deep realizations about each of the Four Noble Truths that together embody the full eradication of ignorance. What was understood by the Buddha on that day of awakening, what constituted his enlightenment, was a deeply penetrating insight into the precise mechanism of our enslavement and the surest means of achieving our liberation. This enlightenment penetrated his mind and transformed his entire being in every conceivable manner:

> This was the insight, understanding, wisdom, knowledge and clarity that arose in me about things I had not been taught.

The *bhavachakra*, in contrast, operates on a cosmic scale, describing the whole matrix in which life itself appears, blossoms, and fades. There must have been great pressure in the centuries after the Buddha's death to give this sermon on the Four Noble Truths the same degree of celestial grandeur. The Sutra of the Turning of the Wheel of the Dharma, however, is sublimely intimate. Its focus is on the transformation of the *inner* life.

It is true that the sutras on the theme of dependent origination share some terminology with the present sutra (specifically: birth, old age, death, and thirst or craving). But the genius of the Twelvefold

Chain of Causation lies in the specificity with which it links each cause and effect. If the Sutra of the Turning of the Wheel of the Dharma is to be considered an abbreviated form of *paticca-samuppada*, as is often claimed, no reason is given why the Buddha did not clarify the matter here. As you can see, the sutra is less than three pages long. The Buddha could have (in fact he certainly *would* have) included the "missing" parts, if indeed there were any.

There is a second line of reasoning that may indicate an additional cause for the drift in the meaning of the term *dukkha-samudaya* over the course of the centuries. The definitions provided within the sutra itself seem clearly to indicate that the second noble truth, *samudaya*, is the *response* to the afflictions listed in the first noble truth, *dukkha*. *Response* may actually be too neutral, too weak a term; the thorny problem indicated here is really the mind's *reactivity*. For the mind's unbridled reactivity is without a doubt what the Buddha understood to be the *cause* of our suffering. The *containment* of that reactivity, through mindfulness practice, is the cause of happiness and ease—the settling of the mind into a state of *samadhi* that can be maintained even in the face of life's unavoidable afflictions. Therefore, although it is perfectly correct to maintain that *samudaya is* the *cause* of our suffering, the *term "samudaya"* itself does not *mean* "the cause of suffering." The *term "samudaya" means* "response"—or reactivity— to afflictions. If this paragraph, upon first reading, has left you a bit confused, then I have successfully demonstrated my point. Confusion is possible. Would you call this strike two against the orthodox view?

[*Enter Mara, "the Bad One."*]

What would seem at first glance to be a rather fanciful recounting of Gautama's awakening appears in a later legend. It presents Gautama, on the eve of his Enlightenment, in a heroic epic struggle with the forces of evil, who are intent upon keeping him from attaining his goal of liberation.

When placed alongside the Sutra of the Turning of the Wheel of the Dharma, however, this legend takes on a new significance. If one views this as an allegorical reenactment, perhaps designed by the Buddha himself or his early followers as an entertainment for a lay audience, we see that in every detail, specifically in the chronology of its scenes, there is a direct parallelism to the First Sermon. Verse for verse, line for line, all the entrances and exits of each of the cast of characters corresponds to David Brazier's interpretation of this sutra and *not* to the orthodox interpretation.

Gautama enters the scene a completely bedraggled wreck of a man. He is in the midst of the most profound spiritual crisis of his life. In desperation, he has decided to make his last stand, so to speak, seated under the Bodhi Tree. He has no guarantee of success. He has spent the last six years of his life vainly attempting to liberate himself from suffering. When he seats himself under the tree, he is the personification of *Dukkha*.

> *My body reached a state of extreme emaciation. Because of eating so little my limbs became like the jointed stems of creepers or bamboo; my backside became like a buffalo's hoof; my backbone, bent or straight, was like corded beads; my jutting ribs the broken rafters of an old house; the gleam of my eyes sunk deep in their sockets was like the gleam of water seen deep down at the bottom of a deep well.*

What *follows* from this *dukkha* (not what *causes* it) is the arrival of Mara. Mara is the personification of *Samudaya*. The temptations offered by Mara are actually the churnings of Gautama's own mind—the *response* to his *afflictions*. They are a catalogue of very realistic possibilities still open to the prince, prompted to appear before his mind's eye by his own depression and disappointment and by the extremity of his afflictions. He could have easily given up this struggle, and he knew it. He could have returned home to the loving embrace of his family, to the security and power of the kingship that was his birthright. There awaited him all of the extravagant pleasures and sensualities of courtly life. This offer of worldly power represents Mara as the "thirst for self re-creation which is associated with greed."

Mara's vast armies arrive next on stage, hideous beings representing all the ancient, twisted energies of habitual responses to affliction: lust, aversion, hunger, thirst, craving, sloth and torpor, cowardice, doubt, hypocrisy, stupidity, false glory, and conceit. Each one of these must have been entertained by the dispirited and exhausted Gautama.

Knowing that *samudaya* "lights upon whatever pleasures are to be found here and there," Mara next sends in his three beautiful daughters to tempt the Buddha: Tanha (thirst for sense pleasure), Raga (craving for being), and Arati (aversion, or craving for not being).

What follows is a beautifully crafted piece of allegory. Mara wants Gautama's throne. In other words, he claims the right to continue to rule Gautama's life—as he always has done—through the power of *samudaya*. But the Lotus Throne of the Buddha's Blissful *Samadhi* is not for sale at any price. In the gesture that initiates for all time the defining practice of the Buddhadharma, Gautama secures his title of the Awakened One by reaching down with his right hand and touching the Earth, which bears witness to his rightful ownership of that exalted state with a powerful roll of thunder, as the Earth quakes and Mara vanishes.

The Earth Touching Gesture (the *bhumisparsha mudra*) is by far the most common representation in art of the Buddha's enlightenment. But what does it signify?

I believe that the meaning of this gesture is contained within the third and fourth noble truths. I also believe that the third and fourth noble truths, in parallel with the first and second, are to be read as an agglutinative combination form: *nirodha-marga*.

NIRODHA-MARGA

The etymological roots of the term used for the third noble truth—*nirodha*—are *ni-*, meaning "down," and *rodha*, the word for an "earthen embankment." The image is of something being contained and protected, as a fire is within a fire pit. The term, as defined in the sutra, clearly denotes a *practice*, that of containing the fiery energies of our thirst for escape that hound us in response to our afflictions. The sutra reads:

> *The noble truth of nirodha, containment, is this: it is the complete capturing of that thirst. It is to let go of, be liberated from and refuse to dwell in the object of that thirst.*

225

Nirodha turns out to be the crux of the Buddha's practice. It is where the distinction is made between what needs to be *contained*—the thirst for relief that arises quite naturally as a response to an affliction—and what needs to be *released*—the object of that thirst, which can only offer temporary relief at best. At worst it may capture one in a whirlpool of craving, clinging, attachment, and disdain, a downward spiral of endless, insatiable thirsts for unsatisfying objects.

This is where the Buddha's insight seems so amazingly modern, displaying the most impeccable logic and scientific acumen. Twenty-three centuries before the publication of Charles Darwin's *The Origin of Species*, the Buddha proposed a formulation entirely compatible with the future theory of natural selection. The brain and body have been programmed by hundreds of millions of years of struggle with afflictions. One of the main functions of intelligence in animals is to generate viable solutions to life-threatening problems as they arise. In Buddhist iconography, the hub of the wheel of *samsara* (cyclical existence) is occupied by three animals, representing greed, hate, and delusion. Each one of our ancestors, in its struggle to acquire food and mates, to flee predators and fight off enemies, has done so from within the deluded conviction of being a separately existing *self* that is identical with the body. The incurved presentation of all of the sense modalities and the urgency and immediacy of our emotional states bolster the intrinsic and entrenched corporeal philosophy of separate self-existence and conspire to bind us ever more tightly in self-serving patterns of reactivity.

The Buddha's practice is to contain the thirst of *samudaya*—to acknowledge its arising as a natural part of the body-mind experience but to release, let go of, and be liberated from the *object* of that thirst, which is an unreal fabrication of mind, a false refuge, and a dream.

Clearly this containment must occur somehow within the body-mind itself. In order to reference the body-mind in detail, in order to address the wide range of afflictions and the infinite varieties of mental and emotional responses, a map of the interior landscape is needed. Later in his ministry, the Buddha would adopt a more neutral map of "the eighteen spheres of mentation" in this and similar contexts (the six sense organs: eyes, ears, nose, tongue, body, mind; the six sense objects: color, sound, smell, taste, touch, object of mind; and the six associated realms of perception).

In the First Sermon, however, I contend that the Buddha was referencing the energetic body-mind map most familiar to this particular audience of ascetic yogis—the classical yogic chakra system.

Before continuing in this vein, let us examine a striking enigma in the orthodox understanding of the fourth noble truth. Each of the four terms—*dukkha, samudaya, nirodha,* and *marga*—is precisely defined in a glossary furnished by the Buddha himself. This list of terms and definitions effectively constitutes the very core of the sermon. Clearly, each of these terms was intended to be viewed as a new coinage, each stamped with the Buddha's own regal imprimatur. He chose what must have seemed to him words whose etymologies were self-evident and therefore impervious to corruption. He further went on to define each of the terms fully and unambiguously in relation to this new Dharma. Despite his best efforts, however, error crept in.

For no sound reason, the fourth noble truth has been understood to differ structurally from the other three. We speak of the Four Noble Truths and the Eightfold Path almost as if they were separate teachings, the latter embedded in the former. In the orthodox view, it is unclear how the eight defining terms are laid out as a *path*, although the word *marga* means precisely "path." Eight specific terms are listed, but there is no real sense of the first term being prerequisite to or leading to the second term, etc., nor is the final term regarded as the goal. Rather, the Eightfold Path is thought of as a set of eight distinct areas of practice to be taken up *as one walks the spiritual path*. In the orthodox view, this is the path toward a future perfection, one that will culminate in the cessation of craving and thence the end of suffering.

Nor does David Brazier upset this view of *marga* in any fundamental sense. In his book, he devotes a chapter to each of the eight "limbs" of the path, of course reframing them in the light of his new translation. In order to understand more fully the significance of the term *marga* and its relationship to *nirodha*, we must consider the setting of this sutra—and the audience.

For six years Gautama had been studying and practicing yogic austerities, most recently with a group of five of his friends, the ascetics mentioned in the sutra. This group included Kondañña, the only one of the five to have immediately awakened upon hearing the words of the Buddha's First Sermon. Gautama had been the disciple of Kamala Alada and, later, Udraka Ramaputra. Although we do not have precise or reliable details of the particular austerities they engaged in, one may assume that the goal of these practices was similar to that of all other esoteric yogic practices: namely, the awakening and development in the aspirant yogi of dormant kundalini energy that rises from the base of the spine through the seven chakras—culminating in the blissful union with Brahma and the attainment of cosmic consciousness at the level of the crown chakra, the thousand-petaled lotus.

The Buddha begins the First Sermon with a direct appeal to the five ascetics, gently chiding them by comparing their pursuit of the lofty goal of escape from the earthly coil through self-mortification to the overindulgent pleasure-seeking that is "the way of ordinary folk." Both strategies he deems "unworthy," "ignoble," and "not conducive to the real purpose of life." He proposes the Middle Way of Right View, Right Thought, Right Speech, Right Action, Right Livelihood, Right Effort, Right Mindfulness, and Right Samadhi. The Pali term here translated as "right" conveys a further sense of "whole" or "complete," as well as (importantly) "all moving in the same direction."

Combining these bits of evidence—the Earth Touching Gesture with its thunderous energetic release; the downwardly directed containment of *nirodha*; the lack of any further clarification within the sutra for what constitute Right View, Right Thought, Right Speech, Right Action, etc.; the immediate comprehension by Kondañña of the Buddha's meaning despite this apparent lack of clarification; an easy double entendre on the Middle Way as the pathway down the middle of the body—the spine and the chakras; the fact that one of the hallmark skillful means of the Buddha was his ability to tailor his message to his audience and that this audience was composed entirely of accomplished yogis; that the *dharma-chakra* (the "wheel of the law" from the title of this sutra) in yogic symbology is the golden ring of the light of consciousness spinning on the index finger of Vishnu's upper right hand, which is said to animate and harmonize all the chakras of the body; that the word for "turning" in the title of the sutra also means "to flow [downward], as water"; and that the list *just happens* to be in the correct sequence—let me propose the following possibility.

Nirodha-marga is the "downward containment path" through the seven chakras, from the crown down to the base—a path that grounds and contains the upward-flaring energy of escape from affliction. The goal of this path, the eighth limb, Right *Samadhi*, is represented by the Lotus Throne upon which the Buddha sits. With this fundamental layer of meaning revealed, the Four Noble Truths can be understood to function as a single, agglutinative sequence: *dukkha-samudaya-nirodha-marga*, which translates (from back to front, to accommodate the grammatical structure of English) as: "the noble, embodied path of containment of the reactive mind in flight from the afflictions of life." From this new vantage point, the sutra is seen to offer direct instruction for the foundational Buddhist practice of mindfulness meditation.

Marga is indeed a path, and Right *Samadhi* is its goal. *Samadhi*, from the Sanskrit for "union" or "putting together," signifies different things to Hindus, Jains, and Buddhists, and within Buddhism itself there are significant discrepancies in its definition. But the *location* of the Buddha's *samadhi* is crystal-clear. It is not "up and out" in some future or distant heaven. He touched the *earth* under the Bodhi Tree at Bodh Gaya. It was the spirits of the *earth* who cried out in gratitude for this teaching in the Deer Park at Isipatana—their delight resounding throughout the heavenly realms. The *Samadhi* of the Buddha marks a fundamental change in our relationship to the world.

Accordingly, when Kondañña later appears in the Surangama Sutra and is asked to describe his entrance into *samadhi* to the assembled bodhisattvas and mahasattvas, he speaks of this moment in the Deer Park as an experience that unified his mind, his body, and the sensory world: "... the thoroughly perfect accommodation, unification, and harmonization of the eighteen spheres of mentation in contact with objects through the sense organs." Wordy, yes, but pure Zen!

It is important not to mistake the map for the territory. Invoking the chakra system is simply a way of stating that since every area of human concern, every level of consciousness, each thought and each emotion must resonate *somewhere* within the body-mind, they must dwell within the natural purview of awareness and are therefore accessible to practice and subject to transformation. In ancient India, the chakra system was not used exclusively by yogis practicing esoteric disciplines. It was the anatomy chart on the wall of every ayurvedic physician's office. It was the common way of understanding how the subtle energies of thought and emotion circulate within the body. This formulation is no more nor less esoteric or exoteric than the mysteriously mundane body-mind itself.

If this had been ancient China, perhaps the Buddha, the Great Physician, would have spoken in terms of meridians and *chi*—if ancient Greece, the four humors. For David Brazier, the reframing of this teaching into language appropriate for our psychological age has been a wellspring for the creative integration of Buddhist spirituality and mental health.

Certainly the Eightfold Path encompasses the orthodox understanding of a defined set of specific areas of focus for spiritual growth. But the Buddha's goal in that teaching moment clearly involved coaxing these emaciated ascetic yogis back into a more "normalized" spirituality by providing a stable energetic bridge back into this world. Each of the words he chose to reference a particular chakra is a humble, grounding term. Thus he revealed *nirvana* within *samsara*, as the awakened life of service, compassion, happiness, and peace.

In each moment, as a specific *samudaya*—a thought of escape—appears in one's body-mind, one is instructed to capture that energy within the chakra from which it originated. One owns the thirst, but releases the object of that thirst. If a reactive response comes in the form of an impulse toward wrong speech, one is to contain that impulse gently within the energetic center of Right Speech, the throat chakra. Should the primitive survival instincts of greed, hate, and delusion arise from the base charka to disturb one's meditation, let those lustful impulses be contained and their unsatisfactory and illusory objects be released through the practice of Right Mindfulness. Errant or unwholesome thoughts are pacified at their point of origin, "the third eye," the sixth charka of Right Thought (or Right Intention).

Right View—of the cosmic truth of this universal Dharma—manifests and stabilizes in the seventh charka, the thousand-petaled lotus crown. Right Livelihood—which addresses the many questions of conscience we wrestle with in procuring our daily bread—settles in the belly, the third chakra. Right Action—the arms and hands being associated with the fourth charka, the heart charka (here one may picture Quan Yin with her thousand outstretched arms)—should be action propelled by love, compassion, equanimity, and sympathetic joy. Finally, let the creative energies of the second chakra not be entirely spent in the mad pursuit of sensual pleasure, but do apply at least *some* portion of that energy toward Right Effort—*upon the Path!*

THE SUTRA OF THE TURNING OF THE WHEEL OF THE DHARMA

Thus have I heard. Once "the One who Enjoys the Spoils of Victory" was staying at Isipatana near Benares. He spoke to the group of five ascetics as follows: Monks, there are two extremes that one who has left the household life should not resort to.

What are they? One is devotion to sense desire and sense pleasure. It is demeaning. It is the way of ordinary folk. It is unworthy and unprofitable. The other is devotion to self-mortification. It is painful and ignoble. It is not conducive to the real purpose of life. Giving up these extremes, the "One who has Been There" has woken up to the Middle Way, which provides insight and understanding and causes peace, wisdom, enlightenment and Nirvana.

The Middle Way is the noble eight-limb way of right view, right thought, right speech, right action, right livelihood, right effort, right mindfulness, and right *samadhi*.

The noble truth of *dukkha*, affliction, is this: birth, old age, sickness, death, grief, lamentation, pain, depression, and agitation are *dukkha*. *Dukkha* is being associated with what you do not like, being separated from what you do like, and not being able to get what you want. In short, the five aggregates of grasping are *dukkha*.

The noble truth of *samudaya*, response to affliction, is this: it is thirst for self re-creation, which is associated with greed. It lights upon whatever pleasures are to be found here and there. It is thirst for sense pleasure, for being, and not being.

The noble truth of *nirodha*, containment, is this: it is the complete capturing of that thirst. It is to let go of, be liberated from and refuse to dwell in the object of that thirst.

The noble truth of *marga*, the right path, is this: it is the noble eight-limb way, namely right view, right thought, right speech, right action, right livelihood, right effort, right mindfulness and right *samadhi*.

"This is the noble truth of affliction"—this was the insight, understanding, wisdom, knowledge and clarity that arose in me about things I had not been taught.

"Affliction should be understood to be a noble truth"—this was the insight, understanding, wisdom, knowledge and clarity that arose in me about things I had not been taught.

"Full understanding of affliction as a noble truth has dawned"—this was the insight, understanding, wisdom, knowledge and clarity that arose in me about things I had not been taught.

"This is the noble truth of response"—this was the insight, understanding, wisdom, knowledge and clarity that arose in me about things I had not been taught.

"Response should be understood to be a noble truth"—this was the insight, understanding, wisdom, knowledge and clarity that arose in me about things I had not been taught.

"Full understanding of response as a noble truth has dawned"—this was the insight, understanding, wisdom, knowledge and clarity that arose in me about things I had not been taught.

"This is the noble truth of containment"—this was the insight, understanding, wisdom, knowledge and clarity that arose in me about things I had not been taught.

"Containment should be understood to be a noble truth"—this was the insight, understanding, wisdom, knowledge and clarity that arose in me about things I had not been taught.

"Full understanding of containment as a noble truth has dawned"—this was the insight, understanding, wisdom, knowledge and clarity that arose in me about things I had not been taught.

"This is the noble truth of the Path"—this was the insight, understanding, wisdom, knowledge and clarity that arose in me about things I had not been taught.

"The Path should be understood to be a noble truth"—this was the insight, understanding, wisdom, knowledge and clarity that arose in me about things I had not been taught.

"Full understanding of the Path as a noble truth has dawned"—this was the insight, understanding, wisdom, knowledge and clarity that arose in me about things I had not been taught.

As long as I had not got a completely clear insight and understanding in all these three ways about each of these Four Noble

Truths, I could not be sure that there was anyone in the world, divine or human, who had woken up to the highest and most complete enlightenment.

However, when my insight and understanding had become completely clear in all these twelve turnings of the wheel, then I knew for sure that there was someone in the world who had woken up to the highest and most complete enlightenment. Then I knew that the liberation of my mind was unassailable.

This is the last step. There is no further step.

When the Victorious One had said this, the five monks were filled with joy. In one of them, Kondañña, the pure Dharma Eye was completely opened. He saw that whatever can arise can be contained.

When the Victorious One had turned the wheel of the Dharma in this way, the spirits of the earth cried out: Near Benares, in the Deer Park at Isipatana, the wheel of the highest Dharma has been turned, and it cannot now be turned back by anyone, human or divine, anywhere in the world.

This cry resounded throughout the heavenly realms. The earth shook. An immeasurable light was now released into the world.

Then the Blessed One said: Venerable Kondañña has understood. And from that day on he was given the name "He Who Understood."

Samyutta Nikaya LVI.11

A SUMMARY OF CHAPTER EIGHTEEN

David Brazier's radically unorthodox translation of the first sermon of the Buddha—The Turning of the Wheel of the Dharma Sutra—sheds "new light" upon the Buddha's enlightenment experience and in so doing shifts the very foundation stones of Buddhist philosophy. What we give up in the orthodox version's cosmic grandeur, we gain back a hundredfold in this retranslation's deeply insightful and pragmatic rationality.

In the orthodox view of the Four Noble Truths, suffering is caused by desire. Suffering can be brought to an end through the complete extinction of desire. This will stop the cycle of reincarnation—for one who accomplishes this lofty goal, there will be no further rebirth.

By reexamining the etymologies of a few key terms, a whole new teaching is brought out in Brazier's translation. I argue that this teaching is so perfectly realized, so all-encompassing, so rational and so scientifically insightful that it must be expressive of the Buddha's original intent. The sutra describes the four universally true and essential realizations that have the capacity to unshackle us from the bonds of illusion. This teaching stabilizes us in a powerful practice, allowing us to meet with genuine serenity the inevitable onslaught of desires and aversions that otherwise plague the inner life of a sentient being.

The following paragraph from the chapter sums up the essential point:

> With this fundamental layer of meaning revealed, the Four Noble Truths are understood to function as a single agglutinative sequence: *dukkha-samudaya-nirodha-marga*, which can be loosely translated (from back to front, to better accommodate the English grammar): "the noble, embodied path of containment of the reactive mind in flight from the afflictions of life." From this new vantage point, the sutra is seen to offer direct instruction for the foundational Buddhist practice of mindfulness meditation.

CHAPTER NINETEEN

BIRTH-BREATH-DEATH

Consciousness is rather uncooperative as an object of scientific study. Science can only study what it can observe or measure, and it can only measure or observe what consciousness *does*—how it reacts when poked or prodded or otherwise enticed into generating a quantifiable response. Yet consciousness only reveals its true, fundamental nature in moments of silence and nonaction.

Buddhism, Taoism, Sufism, Advaita Vedanta, and Mystical Christianity—these traditions all agree that the unbridled expression of the egoic *will* represents the greatest obstacle on the path to awakening to the pure essence of consciousness. It is through the *release* of this egoic will—in the practice of Buddhist mindfulness meditation, in merging with the Tao, in the Sufi's ecstatic union with the Beloved, in the recognition of the deep stillness of Advaita, or in exemplifying the mystical Christian prayer: "Not my will but Thy will be done"—that consciousness comes to the realization of its essential nature, as the *experiencer* of the universe and the willing participant, but not the willful *doer* of it.

Yet from the ego's perspective, hovering above the surface, these practices all look rather suspicious. It is entirely counterintuitive for the ego to go about solving *any* problem by simply letting go; and when the "problem" is the ego itself, there is really not much that *it* can do to help—although it will certainly keep on trying, for that is the primary function of the egoic mechanism.

In the early stages of mindfulness practice, we are instructed simply to note the relentless torrent of longings, disdainful judgments and discomforts emanating from the egoic mind and the physical body. By remaining still in the presence of these disquieting thoughts and emotions we begin to take back proprietorship of our experience. Eventually it becomes clear to us that there is no profit in maintaining the habits of the grasping mind and the trumped up emotional reactivity that distort our perceptions of reality. With time, new habits of openness and authenticity in our thoughts and feelings will supersede the old.

Learning to meditate is a lot like learning to ride a bicycle. No one seems able to tell us the secret to maintaining our centeredness, because few people consciously realize how they do it. Most of us learned to ride a bicycle by *falling*. Eventually the *body* learns to *avoid* falling by making subtle adjustments with the handlebars. The movement we learn to make is completely counterintuitive. If we wish to turn *right*, in order to avoid an obstacle on our left, we must first turn the handlebars to the *left!* The momentum of our body continues to travel forward in a straight line, while the seat and the rear portion of bike follow the front wheel as it veers off toward the left. With the bicycle no longer directly beneath us, gravity brings us slightly lower toward the ground. Hunkered down with our weight now shifted to the *right* side of the bicycle, we are able to make a sharp turn to the right, which avoids the obstacle and centers our weight once again directly over the frame. Being unaware of this complication, we are more likely than not to crash into an object that appears suddenly in our path. We don't have time *not* to think about it. The reactive mind jumps in to "fix" the situation, and deciding quite logically to turn the handlebars *away* from the menacing object, our well-intentioned mind invariably sends us hurtling directly into it.

The art of meditation begins with the development of a deeper, seemingly *counterintuitive intuition*, one that slowly undermines the strategies of the reactive mind and the guarded heart. We are given a peculiar instruction—to attend to the breath—although, in reality, nothing new or exciting is likely to happen there. We are told not to chase after our thoughts and feelings or to indulge them in any way, even though they may in the moment seem quite intriguing and worthy of pursuit. We learn how to remain centered within our experience, present in a state of bare awareness as thoughts and feelings arise, persist for a time, and fade away. What is at first caught only in glimpses—in unexpected moments of clarity and kindness—eventually resolves into the abiding peace of a tranquil mind and loving heart, the unity of wisdom and compassion.

> ...when even just one person, at one time, sits in zazen, they become, imperceptibly, one with each and all of the myriad things and permeate completely all time, so that within the limitless universe, throughout past, future, and present, they are performing the eternal and ceaseless work of guiding beings to enlightenment.
>
> Eihei Dogen Zenji
> *Jijuyu Zammai*
> (Self-Fulfilling Samadhi)

Breath meditation was a technique highly regarded by the Buddha and is the subject of the well-known *Anapanasati Sutta*, the *Sutra on the Full Awareness of Breathing*. The breath appears to be a perfect object of meditation for a variety of impeccably logical reasons, *none* of which, I hasten to add, should be borne in mind while meditating! Such contemplations can be fruitful in one's off-hours, though—an effective way to persuade oneself to settle more easily and without undue resistance into the meditative state. My own analytical mind now seems much more willing to be brought into submission; having been given the opportunity to think all of this through, it appears to have been thoroughly convinced that relinquishing its control is really for its own good. The paragraphs that follow represent the fruit of some of those off-hours contemplations. Much of it is speculative and not entirely *"canonical"*—it is offered for you to corroborate through your own practice experience.

The breath, like the heartbeat, is ultimately under the command of the autonomic nervous system, although we are able, within bounds, to choose to control our rate of breathing by an act of will—when speaking or singing or when holding our breath underwater. Buddhist meditation differs significantly from traditional yogic breath practices in this respect. Buddhists do not generally "use" the breath *(prana)* in order to "attain" higher states of consciousness. In Buddhist mindfulness meditation practice, the breath is *observed*, and through this careful observation one learns to let go of even the subtlest influences of the will upon the breath, until one settles into an experience of the rhythmic, primal, unwilled yet willing breath of deep meditation.

It is quite natural for one to remain more or less oblivious to one's breathing during the normal course of the day. One generally attends to the breath only when breathing becomes difficult—if one is choking on a pretzel or drowning or gasping for air after exertion. The simple act of placing focused attention upon the breath seems automatically to elicit a heightened vigilance, triggering a set of involuntary survival instincts. Breathing is the most basic and essential function of the body. In every single moment of the several hundred million years since our ancestors first emerged onto dry land, having immediate access to the oxygen in the air has been for them (and is for us) quite literally a matter of life and death. Evolution appears to have developed a failsafe method that engages the mind in a positive feedback loop whenever the subject comes up. When attention is placed upon the breath, it seems to signal the mind to focus even *more* attention upon the breath to investigate whether something might be amiss. The quickest way that I have found to drop into a meditative state is to breathe out completely, as if it were my last breath, and wait to see if the body chooses to breathe back in again. When I first attempted this technique, it would invariably bring on a bit of a *panic* reaction. Now, it simply and immediately engages a more *"panoramic"* attentiveness to the sensations of the present moment.

The function of attention is understood to be regulated by powerful neuronal oscillations between the thalamus and the cerebral cortex that entrain with resonance patterns from one or another of the sensory cortices. The focusing of attention upon the bodily sensations of the breath—particularly upon the fully released out-breath (the moment in the breathing cycle of the body's greatest vulnerability)—calls upon the brain to unfix the mind immediately from its current object of contemplation, whatever that may be. The unconscious, instinctual brain interprets our sudden interest in the breath as an indication that something may be about to threaten our breathing—and redirects the corticothalamic oscillations to entrain with this highest priority somatosensory input—simultaneously issuing a general order for heightened vigilance and focused mindfulness throughout the sensorium.

Traditional meditation manuals do not usually include the instruction to "entrain your corticothalamic oscillations with your respiratory somatosensory perceptual stimuli." The effectiveness of the meditation technique is certainly not dependent upon one's knowledge of this arcane material. Mindfulness meditation is the simple practice of returning one's attention to the breath and to the sensations of the present moment. In this way one is able to calm the "monkey mind." Jakusho Kwong Roshi of the Sonoma Mountain Zen Center expresses his understanding of the practice of Zen meditation with the phrase: "Breath sweeps mind." (I also like my father's more tactical phrasing: "Breath *trumps* mind"—for no errant thought form can possibly compete with such visceral intimations of one's own impending mortality.) In these moments of clarity one is given a foretaste of the experience of full awakening.

The sensations associated with breathing occur in the same locations that we normally project the illusion of the separately existing *self*—within the head, the chest, and the trunk. This ancient practice of breath meditation anticipates one of the basic techniques of behavior modification used by cognitive psychologists today; instead of trying to eradicate a persistent bad habit by direct frontal assault, the client is instructed to engage in another activity that is impossible to perform simultaneously with the first. Over time, the neutrality of the breath becomes a welcome stand-in for the stubborn self-concept, allowing the unskillful aspects of the egoic function to become more transparent and yielding to mindful awareness.

Alterations in rhythm and other perceptible qualities exhibited by the breath allow us to maintain awareness of our emotional state, even during periods of deep meditation. A tightness in the breath, an expansion, a yawn or a silent sigh, a short breath, a long breath, a deep or shallow breath all clearly reflect the passing emotional tides. By attending to the breath we can know that we are remaining fully present to the subtle communications of the body during meditation without having to rehash the storylines that first prompted the emotions or engaging in an intellectual analysis of them. On an energetic level, the breath encounters all seven charkas, ascending and descending, with the expansion of the lungs supporting the further opening of the heart charka, the seat of loving-kindness and compassion.

Breath meditation is so simple. The only instruction one really needs is: "Return to the breath." Why concern oneself with all of these other complications? A similar question could be asked regarding the Buddhadharma. The core realizations are fundamentally so simple. Why trouble one's mind trying to make sense of the rest of its vast and complicated canon?

The Buddha himself, of course, never wrote down a single word of his teachings during his lifetime. It was only after several hundred years had passed and many generations of disciples had come and gone that the Buddhist Canon was committed to writing, during the first century B.C.E. Is it even plausible that the Buddha actually spoke every word that is recorded there? Moreover, it is well documented that the loyal followers of a radical reformer such as the Buddha will often reframe his teachings after his death to accord with conventional views and cultural norms. (*See: The New Testament*, Bks. I-XXVII)

These questions are not idle ones. Buddhism has come to the West, and the West is responding to what it sees and hears with a great deal of enthusiasm. Mixed in with the enthusiasm is some justifiable confusion as to precisely what the Buddha's central message actually was. We are now in the process of sorting through the scriptures and artifacts of material culture that have accompanied these spiritual traditions from their homelands to ours. As unlikely as it may sound, America in the twenty-first century promises to be one of the most vibrant crossroads of Buddhist cultures.

The history of Buddhism is one of change and adaptation; all societies that have encountered Buddhism have ended up changing *it* as much as they have been changed *by* it. In America, then, Buddhism will likely be valued most for its *pragmatism*—its emphasis on those practices, like breath meditation, that can actually "get the job (of enlightenment) done." It will not be too harshly judged for its brave initial attempts at a fundamental science and an all-encompassing cosmology—our own physicists, after all, are in hot pursuit of just such a Grand Unified Theory of Everything.

Buddhism is well-known for its encyclopedic "lists of lists." The fourth of the Four Foundations of Mindfulness, for example, is a list of the Five Dharmas of Which to Be Mindful, which in turn is comprised of five more lists of the Four Noble Truths, the Five Hindrances, the Five Skandhas, the Six Senses, and the Seven Factors of Enlightenment.

The point of all of this list making is in one sense simply to build a sturdy *palace of memory* for keeping track of a great number of interwoven ideas and concepts—an important accommodation to the needs of the largely preliterate society in which the Buddha operated. Not only the lists were committed to memory by the monks and nuns in early monastic communities, but also the long and elaborate sutras that elucidated the context and meaning of the Buddha's teachings to which the lists referred. What these lists represent is a distillation in outline form of the Buddha's ongoing lifelong project: to analyze thoroughly every aspect of the mind and body in order to determine which of the inherent capacities of the human being are immediately conducive to awakening, which are problematic but ultimately redeemable, and which are so tainted that they should be entirely abandoned.

When all of the lists and all of the sutras are boiled down to their essence, a single, penultimate word remains before the *silence* consumes all words and all thoughts. That word is the answer given by the Buddha when he was asked, soon after his enlightenment—by a man who met him on the road and was astonished at his peaceful radiance—whether he was a god, a celestial being, a wizard, or a man. "I am *Awake*," was the Buddha's reply.

The Buddhadharma is a rich compendium of spiritual, psychological, and philosophical knowledge on virtually all subjects relating to the transformational arts, based upon one man's profound metaphysical awakening and bolstered by the best—albeit rudimentary—science that was available in the fifth century B.C.E. It is replete with insights of all kinds, daring leaps of imagination, and thoughtful analyses delivered with skillful rhetoric. However, although the Buddhist Canon is immense, it is neither absolutely all-encompassing nor correct in every detail. The vast treasure trove of knowledge and information that has been accumulating over the course of the last two and a half millennia—that has led to our modern scientific understanding of both the large-scale and small-scale functioning of the physical and energetic universe and of the elaborate mechanisms of evolutionary biology in particular—needs finally to be thoroughly sorted through, to verify whether the *essential* teachings of the Buddha still accord with this new vision of reality.

The question I am posing is whether the Buddhadharma should be viewed as a closed system or an open one. From what I can tell, those sutras that can be authoritatively attributed to the Buddha himself reveal a clear process of growth and development in his curriculum and teaching style, as his *sangha* grew into a larger, more complex community. His offerings became more and more varied as time went on—providing brilliant philosophical arguments to confound the intellectuals; deep heart wisdom to heal wounded hearts; structured practices to gather up scattered minds; and ethical teachings to help build a righteous and sustainable society.

I have to believe that if somehow the Buddha could have continued his ministry on into the modern age, he would have delighted in the study of cutting-edge science (although it is difficult to imagine what the neurobiological sciences would look like if they were restrained by his impeccable bioethical standards). The core teaching that he offered from the beginning—The Turning of the Wheel of the Dharma Sutra, with the doctrine of the Four Noble Truths, the Middle Way—is as true now as ever it was or ever it will be. It is my firm belief that, regardless of whatever new insights the outwardly directed scientific study or the inwardly directed self-study of embodied consciousness may yet reveal about the actual processes that facilitate this interdependence of mind and matter, it will all be held within the compass of the whirling wheel of this primordial Buddhadharma.

The theory of neurobioluminescence in the evolution of sensorium consciousness is offered in the spirit of furthering the dharmic discourse. I fully expect it to raise some eyebrows among those who wish to keep their Dharma contained in a hermetically sealed, tamper-proof box. I'm willing to take a peek inside, like foolish Psyche—or Schrödinger's tenderhearted lab assistant—because I believe the cat is still alive in there and needs some fresh air! Recall that it was only when Pandora opened her box for a second time that the one kindly creature, Hope, was finally released.

It is great good fortune to be born in a human body. It is greater good fortune still if, as a human being, one hears of the Dharma and practices it. The Dharma teaches us a better way of being—calmer, easier going, more loving, compassionate, and fearless. The Dharma is ruthless in pointing out our basic failing—our weak-kneed subservience to our desires and aversions. It is ruthless because it knows for a fact that the reward of awakening is entirely attainable and that to waste this opportunity of a precious human birth is unconscionable. That is why the bodhisattva Quan Yin stays at her post. The joy she imparts in liberating sentient beings is indescribable. The suffering she eradicates is unspeakable.

Contemplating the disturbing reality of the bloody mayhem that has propelled us along our evolutionary path—engaging the mind in this unremitting *"Realmeditazion"*—can be difficult at times, no doubt about it—a challenging alternative to what all of that New Age literature promised would be a lighthearted romp of body and spirit. It seems counterintuitive to pry into these dark closets, hoping to discover there the source of true happiness and liberation. What I am most concerned with is trying to prepare you, gentle reader, psychologically and spiritually, so that you can face this unsettling news about who you actually are and how this befell you—and to remind you of the bliss and freedom that attends your awakening.

Here is a poem I wrote—maybe *it* will cheer you up. Truly, if you can make it through this, you're ready for anything. It's about the sacred, coiling thread of deoxyribonucleic acid (DNA) that wends the way of this human form up from the primeval sea. The word *sutra*, which refers to the sacred Buddhist teachings, also means *"thread"* (from the same Indo-European root as our word *suture*). Like the spool of thread given to the Greek hero Theseus by Ariadne, this living thread must somehow lead us out of the labyrinth of the Minotaur, back into the bright sunshine of freedom. The poem contains an image of the Hindu goddess Kali, often depicted with bloody fangs and a lolling tongue, who takes on both aspects of the creative and the devouring Divine Mother. It ends with an image of Quan Yin—the thousand armed bodhisattva of compassion—a transgendered, *feminine* Chinese adaptation of the *masculine* Indian Avalokiteshvara and Tibetan Chenrezig—who is portrayed with an open eye in the palm of each of her thousand hands, symbolizing her infinite awareness of all suffering beings and her ceaseless vow to save them.

PSYCHE'S PALACE

Within a myriad mothers' wombs, Awakening One,
the blood-red Sutra that named your form
was read—and read again—a hundred billion times—
your eyes and ears chanted forth
from the ancient texts, kept bound in sacred coils.

Bound toward liberation, this changeling body
hobbles through endless time,
wearing—within—a shimmering veil of truth and lies
to first entice and then ensnare a soul.

After each leap into the air of being,
this flying fish splashes down again
in death, into the great womb-sea.

And could you in good conscience seed
a virgin world with this sinister code,
knowing the Awakened One only comes
after a billion years of violent striving
from that scum?

Have you not noticed Quan Yin's thousand outstretched arms?—
her long-kept vows of Great Compassion made
one sleepless night of horror and of bliss,
with the sudden memory,
in the opening of each palm's pure Dharma Eye,
of the multitude of lifetimes she had been,
of those thousand tentacles plucking soft-bodied prey
into her—Kali's—yawning maw.

No eye ever opened upon this world unaccompanied by a claw.

The theory of neurobioluminescence in the evolution of sensorium consciousness offers a serious challenge to even the bravest spiritual seeker. Are we willing to look squarely upon the stark assemblage of irrefutable facts that points out the precarious, transitory nature of our individual body-minds and the rampant, merciless evolutionary rivalry that carved us out in the first place? Can we resist the urge to fall into an existential despair from the realization that these bodies are the handiwork of an inevitable history of murderous competition, chicanery, selfishness, and greed? Can we own at least the possibility that in some former lives we may have participated in this bloodbath in a less than exemplary manner? Is it possible to maintain connection with this difficult knowledge, yet somehow engender a spark of gratitude within our hearts for the opportunity now before us to make amends, to take full advantage of this one precious human birth, to vow with the Buddha and with Quan Yin and with all the other remarkable souls upon this planet to help bring about the radical shift of the great awakening?

Dharma gates are boundless, I vow to enter them.

If these last few pages have seemed a bit "over-the-top"—then you are probably catching on! The human mind was never designed to contemplate *itself*. To attempt to do so sets us spinning—like a child trying to catch a glimpse of the back of his own head. Throw into the mix the Cosmic Void and the "full catastrophe" of four billion years of evolutionary history; generalize your findings to all sentient beings, past, present, and future; devise an appropriate ethics and vow to uphold these principles, so help you *God* (though there may not even *be* one), 'til the last sentient being is saved—all the while contemplating your own mortality—and you're bound sooner or later to blow a gasket.

The mind can only take so much of this. These are profound existential questions, all of them, and will never be tied up into one neat bundle, triumphantly tagged: "Done!" All the same, if you happen to be of a philosophical or scientific bent, these ideas are simply too enticing to be left on the high shelf, out of reach. So we toy with them, toss them back and forth with our friends, and juggle them alone late at night in our easy chair.

The predicament is that the mind is simply not equipped to examine anything entirely *in the abstract*. The contemplation of consciousness *by* consciousness is fundamentally an unfathomable *koan*: it engages the egoic mind and its guiding emotions of desire and aversion in a hunt for a solution—perhaps more rarefied than its usual fare—but nonetheless, the mind doggedly pursues the game. Giving the mind such an intangible and elusive object to track down condemns it to a state of perpetual motion—which may ultimately be experienced as *"high anxiety"* or *existential dread*.

This is where breath meditation becomes an essential practice for all of us "deep thinkers"—a failsafe technique for maintaining proper mental hygiene. The mind is quite a clever inventor, a competent tinker and handyman. It can find a solution to almost any practical problem and was designed by evolution expressly to perform such basic, day-to-day tasks. If you are strolling through the jungle, for instance, and notice a tasty piece of fruit dangling just out of reach, simply ask your mind what to do, and it will correctly suggest that you find yourself a stick with which to whack it down. The mind can successfully handle even more complex problems, but there is a definite limit to the number of plates it can keep spinning without itself spinning out of control.

Meditation interrupts the mind's otherwise endless cycling. Even if you can successfully avoid catching the mind upon the alluring hook of an existential conundrum, there will always be a future for it to worry about and a past for it to regret. Even in the *now*, there are many "elsewheres" that vie for the mind's attention. Breath sweeps the dust of these habitual thought forms from the doorstep, inviting fresh perceptions—sights, sounds, smells, sensations of the body, and novel thoughts and emotions—to enter in for *this* moment—and something entirely new, for the next.

Awakening may be accompanied by a profound shift in one's perceptions—a sudden realization of the "unity of substance" between oneself and what is being presented through the "six sense gates" (*sadayatana*). For it is the integrated activity of the six sense fields that gives the hollow spherical structure to the sensorium surround and opens up sufficient space in the center for an interjected "self-concept" to exist. Mindfulness meditation practice helps rouse Psyche from her waking dream, as she sits dozing upon the central throne that she has occupied for the last several hundred million years of her evolution. This practice allows her to catch a glimpse of who she truly is—to realize the elemental nature of her *awareness*—even in the midst of such powerful transformations of perception.

The brain has evolved a very pleasant mind state into which Psyche is allowed to slip whenever conditions of safety and satiation prevail.

It is the state of *calm abiding*—the mind relaxed, yet alert, eyes softly scanning the horizon, the body in quiet equipoise, but ever ready for action. Conservative of energy, restful, and restorative, this "prototypical meditative state" has evolved into a very strong chaotic attractor within the brains of many species of higher animals. In humans, the alpha brain waves associated with this state (with frequencies ranging from 9 to 14 cycles per second) are slower and higher in amplitude than the beta brain waves that indicate full arousal. The fortuitous combination of these two gifts of evolution—the innate ability of the human mind-brain to come to rest in a state of calm abiding *(samatha)* and the heightened concentration of awareness provided by focused attention upon on the breath *(vipassana)*—blend seamlessly in the practice of mindfulness meditation.

Chan (Zen) Master Wumen Huikai (1183-1260) wrote the following commentary in *The Gateless Gate,* his collection of koans from the ancient masters:

> *All you Zen students training in the Way, don't be victimized by sounds; don't be blinded by forms. You may have realization on hearing a sound or awaken on seeing a form; this is natural. But don't you know that real living Zen students can ride sounds and veil forms? They will see all and sundry clearly. They handle each and every thing deftly. Perhaps you are such a person, but tell me: does the sound come to the ear, or does the ear go to the sound? And if you have transcended sound and silence, what do you say at such a point? If you listen with your ear, it's hard to understand. If you hear with your eye, you are intimate at last.*

This is echoed by Dogen Zenji in his *Genjokoan*:

> *That the self advances and confirms the ten thousand things is called delusion. That the ten thousand things advance and confirm the self is enlightenment.*

There is a hazard associated with some forms of intensive meditation—a good enough reason to place oneself under the guidance of a competent teacher. As the illusion of a separately existing *self* begins to break down and the doors of perception open more widely, there arises the possibility that one will misinterpret this melding of the *self* with the sensory world as evidence of having "attained" some sort of *expansion of consciousness*. Dogen warns against the folly of this in the section of the *Genjokoan* cited above. The danger, of course, is of "spiritual inflation"—in lieu of an actual *dispersion* of the egoic self-concept, the ego is simply made to expand outward until it appears to itself to have "become" the whole universe—a narcissistic projection upon an illusory screen of immense proportions.

To avoid this peril, it is generally recommended that one make a serious commitment to the ethical practices of kindness, generosity, and patience (from the Buddhist list of *paramitas*, or "perfections"). This shifts the focus of one's attention—giving one the opportunity to consider the reality of the subjective experience of another sentient being whose sensorium perspective one is not directly privy to (which is the practice of *compassion*) while decreasing the amount of anxious attention one places upon one's own unreal and selfish *self* (the practice of *wisdom*), until a reasonable balance between *self* and *other*—a paradoxical understanding of the simultaneous reality and unreality of both—is finally achieved.

The theory of neurobioluminescence in the evolution of sensorium consciousness cannot be viewed as an "ethically neutral" theoretical construct. It has a great deal to say about our responsibility to consider the inner reality of our fellow sentient beings, both human and nonhuman, with whom we share this planet's finite surface. If it is indeed true that humans are not the sole proprietors of sensorium consciousness; that all animals share the same fears, longings, and pains that we ourselves experience; that the mouse looks out of his own small eyes upon a world just as large and imposing as the one we see—how can we then maintain that an animal would *not* project what can only be called a *self* into the hollow center of his being in precisely the same way that we do? Perhaps animals don't give that interjected *self* a name—but does that make their experience any less real? Other animals may not worry incessantly about their mortality or their fate in an afterlife as we do (and probably should do, considering the inordinate amount of suffering we as a species routinely inflict upon the animal world). We may never know the intimate details of the inner lives of our animal cousins—at least not in this incarnation—but we can imagine how the view from within may be very much the same as our own.

There are a great number of impressive mental faculties that obviously make human beings special, but do they make us *inordinately* special—categorically different? If it follows logically from first principles that sensorium consciousness can only function in the context of perceived pleasure and pain, and we see cortical structures in animals also arranged for sensorium presentation in precisely the same ways that they are arranged in our own brains—pleasured and pained by the same neuropeptides, the same "molecules of emotion"—surely we can finally and definitively state that we share at least *this* poignant property with all animate life: we all suffer—and we all desire happiness.

That's about all I dare say—for now—on the subject. I'm sure you can find your own way through these ethical thickets. Speaking of thickets and finding one's way, the wildflowers have been stunning this spring, because of all the rain we've been getting. Now that the weather is clearing up, it's well worth a hike up Sugarloaf Mountain. Fill your sensorium with unbridled beauty—for that, too, is the meaning of life. I'm just pointing out the bits of poison oak that I have noticed here and there, gracing the edges of this glorious path.

PSYCHE'S PALACE

A SUMMARY OF CHAPTER NINETEEN

Meandering through the wide territory opened up by this new theoretical construct, a few tentative observations and advisements are made for how to practice meditative awareness, compassion, and joy in this sparkly new world of the bioluminescent brain. We suddenly find ourselves in circumstances where science and spirituality meet and meld. No longer is the brain an alienating biocomputer; now it is seen as the source of resonance and rationality—of beauty and music and pleasure.

The ultimate goal of all spiritual practice is to awaken to that which has always been awake, to discover that it is the seeker himself who is the ultimate object of his own spiritual quest, to realize the eternal unity of the experiencer and all possible experiences. This is not because God likes a good riddle. It is simply a stubborn fact of metaphysics that consciousness happens to be constructed in this peculiar way: being both subject and object of its own experience.

Awakening occurs when the mind settles upon itself without the distinction of perceiver and perceived. By focusing one's attention upon the breath in mindfulness meditation, one enters the state most conducive to such an awakening. Although the moment of enlightenment is impossible to predict—an accident, they say—a solid meditation practice makes one decidedly more accident prone.

The theory of neurobioluminescence in the evolution of sensorium consciousness paints a grim picture of life upon this planet over the course of the last billion years or so. Predation, hunger, pain, fear, lust, and aggression were part of the daily routine of our ancient ancestors. The evolutionary carnage that has somehow brought us to our present situation, a relatively peaceful moment of possible spiritual transcendence, should prompt us to redouble our efforts. How likely is another rebirth in such an auspicious realm?

CHAPTER TWENTY

BOTH SIDES NOW

> *I've looked at clouds from both sides now,*
> *from up and down, and still somehow*
> *it's cloud illusions I recall.*
> *I really don't know clouds at all.*
>
> Joni Mitchell
> *Both Sides Now*

P syche embarked upon her perilous journey many hundreds of millions of years ago, a captive in the hold of a dimly lit vessel, peering through cloudy portholes upon cavernous seascapes, the muffled sounds of passing sea monsters disturbing her half slumber with fitful dreams of rows of razor-sharp teeth.

Now she walks upright and is civilized and learned. She has developed language and culture and science and books, and she practices meditation and speculates with her *Light* about the *Light* within her.

With this final chapter we are coming to the end of a long and meandering path. We have seen a lot of countryside: some familiar landscapes, some new views of places we presumed we knew well, and occasionally some completely startling and unfamiliar sights. Now we'll look back in quick glimpses at the territory we've covered.

The theory of neurobioluminescence in the evolution of sensorium consciousness posits that one of the principal functions of the brain is to *create* this very view of the countryside, the sounds of the birds in the trees, and the feel of the breeze upon the skin, *directly* from the energetic fields produced by the firings of a certain class of presentation neurons. Arrayed in intricate matrices of specialized, *Light*-focusing cortical columns, these presentation neurons carpet the outer surfaces of the sensory cortices of the cerebral cortex. The electrical voltage generated by their firings is not an *epiphenomenon*, but an end in itself.

There is something *shimmering* overlaying the computational grid formed by the meshwork of cerebral cortex neurons, axons, and dendrites. When a presentation neuron fires, it produces a minute, localized electromagnetic field disturbance—a tiny "spark" of *Light*—that, synchronized with the "sparks" from billions of other firing neurons, creates all of the powerful sensuous imagery experienced within the sensorium. Galleries chock-full of carefully crafted somatosensory, auditory, olfactory, gustatory, and visual depictions, artfully displayed upon a wide array of cortical and subcortical presentation structures, continuously vie for our attention.

The hemispheres of the cerebral cortex, unfurled, would be the size of two extra-large pizzas—a pair of contiguous, two-dimensional surfaces begging to be acknowledged for what they self-evidently are: bioluminescent planes of pixellated cortical columns displaying an infinite variety of microscopically detailed patterns within a minutely organized, constantly fluctuating quantum electromagnetic field. The crystal-clear cerebrospinal fluid that coats the surfaces of the cortex and the crystal-clear cerebral interstitial fluid that interpenetrates its neuronal structures form the medium that propagates the *Light* of conscious display. This *Light* is both subject and object, real and illusory, existent yet completely immaterial. The lightbody traverses the hills and valleys of the cerebral cortex, hugging the surfaces of the transparent *pia mater*, receiving its own structural essence from the electromagnetic field produced by the brain of its animal host.

Neural resonance is the mechanism that allows for the presentation of beauty and for all forms of aesthetic evaluation. Aesthetics, broadly defined, can be said to be the most basic faculty possessed by sentient beings, the common characteristic motivating all animate life. Determining what is desirable and what is undesirable is the fundamental purpose of sense organs and sensory cortices. An animal has the choice either to move toward or away from any object or being that it encounters. *"Like"* and *"dislike"* are the two primordial neural resonance modalities that shift the sentient being into forward gear or into reverse. With the quick release of an appropriate batch of neuropeptides, the sensorium experience is shifted from pleasure to pain and back to neutral, from desire to aversion and back to indifference.

Also by means of the neural resonance chamber, the illusion of three-dimensional form is allowed to expand into the dimension of time—permitting the purview of the *self* to extend not only several feet out in all directions but several seconds back into the past as well. A vivid imagination also propels the mind into a probable immediate future. The *self* inflates from the center outward to claim its small territory of personal space and time.

And how did all of this come to be? What began as a nonconscious two-dimensional retinotopic map arranged simply to allow for efficient cybernetic analysis of visual data simply *became*—one day, quite by accident—the vivid presentation screen of the first sentient being. A compelling light show mirroring an otherwise unknowable outside world began rather unceremoniously as an epiphenomenal leakage of electromagnetic energy—from nonconscious neurons firing their bits of computational analysis of objects appearing within their respective sectors of the visual field. Then, as if in a flash of *Light*, the brilliant display came suddenly to life.

Actually, *all* light—every photon, every electron-positron pair that comprises the quantum electromagnetic field—exhibits what seems to be best described as an elemental form of *awareness*. Attractive and repulsive forces are what keep the electromagnetic field in constant motion—apparently even elementary particles know what they like and dislike—being as predictable in their passions as are we in ours! Like a homeless waif taken in by the biophysical organism, light is given a purpose and a destiny—which is to coevolve with the earliest sentient life forms until at last it achieves its autonomy as a *soul*. It becomes the *Light* of consciousness that informs each sentient being.

If Jason W. Brown's theory of *microgenesis* is correct, this sentient being comes newly into being from the ground of being many times each second. We are the fountains of eternal youth, the fresh iterations arising from the *plenum void*, thoroughly briefed on our backstory during the short "*Light*-rail" ride from the brainstem to the cerebral cortex. Herein lies the great value of spiritual introspection and mindfulness meditation practice: one learns how to encounter oneself in full awareness while still in the state of becoming—and not yet entirely committed to a plodding reenactment of the *status quo*.

The *"illusory self"* and the sensorium that makes it happen are certainly the most astonishing artifacts of evolutionary biology. Once sensorium consciousness made its initial appearance in evolutionary history, it soon became mandatory equipment for any creature hoping to compete head-to-head in this brave new world of conscious awareness. What an advantage it must have been to enter the fray equipped with a real, live *Psyche* inside of one's skull—someone who actually cared about her own survival—who could imagine what death might feel like, having experienced over the course of her lifetime the full gamut of pleasures and pains. She would be brave in battle and wise in retreat. And if you are wondering how she fared, just look around the planet—Psyche is everywhere!

"Take this and drink, Psyche, and be immortal. Cupid shall never break free from this knot of love in which he is tied. Your nuptials shall be perpetual." A hundred million years she has spent in the company of her beloved—in blissful union and in blissful ignorance. She gazes through his eyes upon the infinite universe that lines the inward curving of his skull—as Shakti, now, in the sensuous embrace of Shiva.

We are the inheritors of this miraculous body-mind, born at the bleakest and most promising juncture in human history. Yet the way through this passage is not difficult.

The Path is clear and brightly illuminated.

A HAPPILY EVER AFTERWORD

The great way is not difficult
for those who do not pick and choose.
When preferences are cast aside,
the way stands clear and undisguised.
But even slight distinctions made
set earth and heaven far apart.
If you would clearly see the truth,
discard opinions pro and con.
To founder in dislike and like
is nothing but the mind's disease,
and not to see the way's deep truth
disturbs the mind's essential peace.
The way is perfect, like vast space,
where there's no lack and no excess.
Our choice to choose and to reject
prevents our seeing this simple truth.
Both striving for the outer world
as well as for the inner void
condemns us to entangled lives.
Just calmly see that all is one,
and by themselves false views will go.
...
Delusions spawn dualities.
These dreams are naught but "flowers of air"—
why work so hard at grasping them?
Both gain and loss and right and wrong,
once and for all get rid of them.
When you no longer are asleep,
all dreams will vanish by themselves.

If mind does not discriminate,
all things are as they are, as one.
To go to this mysterious source
frees us from all entanglements.
When all is seen with equal mind,
to our self nature we return.

...

With single mind, one with the way,
all ego-centered strivings cease.
Doubts and confusions disappear,
and so true faith pervades our life.
There is no thing that clings to us
and nothing that is left behind.
All's self-revealing, void, and clear,
without exerting power of mind.

...

When faith and mind are not separate,
and not separate are mind and faith,
this is beyond all words, all thought,
for here, there is no yesterday,
no tomorrow,
no today.

Jianzhi Sengcan (d. 606)
Third Chan Patriarch
Xin Xin Ming

By the power and the truth of this practice, may all beings have happiness and the causes of happiness. May all be free from sorrow and the causes of sorrow. May all never be separated from the sacred happiness that is without sorrow. And may all live in equanimity, without too much attachment or too much aversion, and live believing in the equality of all that lives.

INDEX

A

Abhidharma 35, 49, 50, 52-56, 193-94, 199
Advaita 135-36, 235
advantage, body over soul ...38, 80-81, 196, 200
aesthetics..... 9, 25, 27, 31, 40, 68, 198, 210, 213, 253
afflictions (*dukkha*) 216-30
agency
    illusion of.............................72, 184
    proof of the existence of 188-89
algorithms ..169-70, 172, 177, 203, 206
alleles... 110
alpha brain waves 97, 247
altruism 110
amphitheater, LGN as 156
amygdala 188
Anapanasati Sutta 237
anatman/atman................. 45-46, 136
ancestors 16-17, 47, 61, 77, 86, 90, 102, 121, 128-29, 155, 187, 199, 226, 240
    first conscious 77, 121, 123, 128-29, 161, 179-80, 188, 197, 251
angle of orientation, perception of 147, 161-62
angst, existential 40, 245
anima 104, 149
animals, experience of selfhood in . 39, 40, 65, 159, 161, 210
Answers in Genesis web site..............67
arachnoid mater
    insubstantiality of......................163
Aristotle, *camera obscura* 126-27
astroglial cells 164, 165
attention 170-71, 187, 191, 195, 208, 240, 248
attractor basins 98-99
attractors .... See chaotic attractors
auditory cortex... 14, 16, 124, 170, 203, 205-06
Aum 204
automatons 39, 71, 107, 123
aversion ....See desire and aversion
awakening 40, 52, 62, 65-66, 76-77, 217, 221, 235, 239, 244, 246

awareness 11, 14, 26, 29, 30, 32, 34-38, 46, 50, 53, 60, 65-66, 68, 71-73, 76-77, 80-83, 87, 93, 95, 104, 107, 110, 113-14, 118-19, 120, 123, 128-29, 143, 148, 171, 180, 183, 185-91, 194-95, 229, 237, 239, 244, 247, 253-54
    bare 15, 35, 119, 129, 237
    quantum field and.11, 14-15, 29-30, 32, 34, 66, 119, 128, 190, 194, 247, 253
axon hillock 37, 134-35
axon-dendrite connectivity. 3, 5-6, 36-38, 99, 146-48, 169-70, 185

B

bardo 77, 138
Beauty of the Peacock Tail', 'The 67, 70
being, first conscious .... See ancestors, first conscious
being, nonbeing and 136
bewilderment 35, 54-55, 73-74, 76, 81, 128, 175, 185
bhavachakra 108, 123-24
bhumisparsha mudra..................... 225
binary code 7, 13, 206-07
binaural input 12, 124
binding problem, the 7-8, 151
binocular vision.................. 124, 151-55
    conjectural model for............151-56
biocomputer 38, 133, 162, 165, 169, 171
biogenetic law.................................... 46
biophotons...................................... 160
blind spot.. 72
blobs ....See color, blobs
blood-brain barrier 164-65, 188
Bodhisattvas 64, 108, 118, 229, 243
    Four Bodhisattva Vows, The....... 64
body-mind .25, 41, 46, 60, 62, 94, 100, 102, 103, 111, 128-29, 179, 196, 203, 220, 226, 229, 230, 245, 254
body/mind................... See mind/body
Book of Causal Relations, The 199
brain .....See also more specific entries and the illustration on page *xii*

257

INDEX

brain (contd.)
    brainstem....... 35,-38, 41, 47-48, 74
        fetal 184-85
        microgenesis and.. 123, 175, 181-86, 188, 195, 254
    chaotic system, as.6, 97-98, 171-74, 190, 209, 247
    chemistry40, 94-104, 122, 165, 170, 172, 185, 191, 196, 249, 253
    counting and.........................202-03
    cybernetic 197, 123, 125
    development, fetal, consciousness's role in 185-86
    frequency analysis by 3, 202-06
    harmonic analysis by... 31, 124, 174, 202-06
    mammalian, old 35, 47
    many brains theory and 100
    material consistency of 165
    nonlinear system, as.... 98, 190, 199
    parietal lobe 15-17, 48, 84, 166
    passive medium, as148, 169, 172-75
    principal function of......4, 226, 252
    reptilian 35, 47, 185
    reverse orientation of15-17, 86, 254
    states97, 100-02, 122, 129, 185, 210
    topology of... 12, 18, 33, 42, 83, 124
brain waves 97, 247
brain-space 147-48
Brazier, David... 215-20, 223, 227, 229
    analysis of the Turning of the
        Wheel of the Dharma Sutra 215-33
breath meditation.............237-40, 246
Brown, Jason W. 35-37, 41, 47-49, 141, 180-83, 193, 254
Bruno, Giordano............................ 208
Buber, Martin 82
Buddha's First Sermon, The 215-33
Buddhadharma.....45, 56, 225, 240-42
Buddhism 136, 237
    Brazier, David and................215-33
    canon of 215, 221, 237, 240-41
    Four Noble Truths, The..62, 215-33
    Mahayana 53, 56, 63-64, 135-36
        mind, view of 53, 57-58
    orthodox view of....63, 216-19, 223, 226-27, 230
    practice of perfections and........248
    practitioners of 106
    sentient beings in40, 45-46, 61, 64-66, 71-74, 107, 109, 118, 136, 159, 220, 243, 245, 248, 253-54
    teachings of45, 55, 56, 108-09, 135, 194, 215
    Vajrayana (Tibetan).......49, 55, 108
    Zen ... 43, 64, 97, 108, 111, 135, 237, 239, 247
Buddhist "Rosetta Stone"............... 215
Buddhist psychology *See Abhidharma*
Burgess, Stuart 67, 69

C

calm abiding 247
calories consumed by brain..... 133-34, 189
camera obscura...........................126-27
Candide 175-76
carnivores 113-14, 122
causality, *karma* and............ 57, 59-62
cerebral cortex 83, 144-46, 160-66, 206, 210, 252, 254
    two-dimensional surface, as.......33, 124-25, 144, 153, 162, 252-53
cerebral interstitial fluid 12, 164-65, 252
cerebral multiplex........................... 126
cerebrospinal fluid.... 12, 144, 163, 165, 188, 252
    archaic brain structures, in 188
    clarity of 12, 163-65, 188, 252
    sensorium, role in...................... 163
    shock absorber, as 163, 165
channels, voltage-gated sodium ion .3, 33, 134
Chaos Theory 98, 173, 190
chaotic
    attractors.................97-98, 173, 247
    resonators 7, 97-98, 172-74, 211
    systems...................... 173, 190, 209
chase lights, theatrical..................2, 38
chemical era......................................112
chiaroscuro 154
chords, musical...3, 7, 95, 97, 101, 124, 173, 205
chromatophores..........................68-69
circuits, computer.. 3, 4, 11, 13, 31, 121, 171, 189, 206
cochlear nucleus 124
Cogito ergo sum 189

258

INDEX

color
- blobs 147, 159-60
- cortical columns and..... 13, 145, 47, 159-62, 165, 252
- perception of..............................148

color constancy 147
columnar organization 146, 162
community of mind 176, 197, 200
compassion . 34, 57, 61, 64, 67, 82, 99, 136, 216, 230, 237, 240, 243, 244, 248
competition, genetic 110
complexification 120, 171, 181, 188
computer paradigm 1, 4, 6-7, 13, 31, 38, 40, 81, 121, 133, 136, 149, 152, 162, 165, 169, 206-07
computer/resonator issue 7
conduit between worlds............. 20, 45
confusion...................... 54, 67, 77, 240
connectivity, neuronal, two types .. 147
consciousness... 1-8, 29-39, 48-56, 66-67, 76-77, 86-90, 95, 101, 113-14, 119-23, 127-29, 137-39, 141-44, 148-49, 155, 157-61, 169-74, 179-91, 195-200, 207, 210-15, 228, 235, 242-49, 254
- animals and................. 65, 107, 248
- Buddhist view of 157
- cybernetic animals without . 39, 125
- eight modes of............................. 54
- elsewhere in the universe 129
- evolution of 184, 120-21, 210
- function of.................................. 190
- information processing, contrasted with................................... 168-74
- location of............................. 143-44
- not within neurons 148
- planes of 180, 182, 186, 194
- six senses and...54-55, 83, 110, 128, 137, 202, 209, 226, 241, 246
- storehouse 59
- universal..................................... 118
- viability during development 187

Containment of the Response to Afflictions, the...........................218
contiguity 201
coordinate system, Cartesian model of consciousness 178
cortex, cortices...*See* specific (auditory, visual, etc.)

cortical columns .. 13, 145-46, 150, 153, 157, 159-62, 165-66, 208, 252
cosmology . 53, 56, 57, 58, 62, 216, 240
counting, time and 202-03
courage 26, 80
creationism....... 56, 57, 70, 71, 118, 119
cube of consciousness 178, 180
Cupid and Psyche......................93, 101
- Myth of 19-26, 77-80
- relationship between... 24, 101, 165, 254

curiosity, Psyche's 26
cuttlefish 68-69
cybernetic era 113-14
cybernetics 7, 40, 65, 107, 120-23, 125-28, 161, 189, 197, 253
cyclical existence 25, 108, 226
cytoarchitecture ..38, 123, 125, 136-37, 156-57, 164, 187

D

Dalai Lama 49, 60
Darwin, Charles.... 44, 61, 117, 119, 226
Darwinism.............. 57, 70, 91, 117, 119
- challenge to Buddhism 57
- creationism and.......................... 119

data, sense .. 4, 6, 31, 71, 80-81, 83-84, 89, 103, 124-28, 158, 173, 195, 197, 203, 207, 211, 253
Dawkins, Richard............................ 110
death ..20, 25, 45, 54, 79, 80, 102, 108, 110, 137-38, 186, 196, 218, 220-22, 238, 244, 254
Deep Blue 7
Deer Park at Isipatana 229
delta brain waves 97
delusion .25, 40, 46, 59, 60, 64-67, 70, 107-09, 112, 198, 226, 230, 248, 255
dendrites.....3, 5, 6, 36-37, 146-49, 169
dependent origination 44, 57, 217, 221
depth of field, perception of .. 147, 153, 166
Descartes, René....... 1, 8, 135, 149, 189
desire and aversion 25, 35, 40, 55, 189, 197-98, 216, 224, 234, 243, 253
destiny 191, 253
development
- consciousness, role in 184-86
- ego... 53-55

259

INDEX

development (contd.)
   fetal 35-38, 46-48
   model, cube of consciousness ... 178
Dhammapada 58
dharma-chakra............................228
display....5-6, 12-16, 30, 36, 38, 66-72, 82, 84-85, 89, 97-98, 121, 126-29, 132-34, 137, 145, 148-49, 151, 153, 157-62, 165, 170, 177, 202-07, 211, 252-53
   mental realm, of 12, 16, 157, 207
divine intervention 119
DNA 5, 60-61, 109-12, 117, 194, 243
Dogen Zenji 111, 237, 247-48
dots, phosphorescent 6
doubled images 152-54
dream consciousness 37, 180-83
dualism, Cartesian....................... 2, 163
duality, mind/body 36, 136, 174-76
dukkha 218-33
dukkha-samudaya-nirodha-marga 229
dura mater 163
Dzogchen 135-36

E

Earth Touching Gesture, the..........228
edges, perception of 125, 147, 149
efficiency.............125, 128, 133-34, 199
ego....44, 50, 66, 72, 111, 136, 176, 196, 235-36, 248, 256
   Abhidharma, in 53-55
   development of........................53-55
   function of 236
   inflation of 248
egogenesis................................. 53, 193
egolessness 53
eighteen spheres of mentation......226, 229
Eightfold Path, the . 215, 219, 227, 230
electroencephalogram.................. 5, 97
electromagnetic field 3, 5-6, 12, 29-37, 45, 71, 83, 130, 134-35, 156, 165, 171, 174, 183, 190, 202, 252-53
   lightbody and.. 30, 36-37, 136, 148-49, 151-63, 171-75, 181-84, 187-90, 252
   microscopic intricacy of 5, 252
electromagnetism..30, 32, 39, 89, 135, 174, 186, 204, 207

epiphenomenal leakage of.........253
   structured 207
elephant and the blind men 8, 65
embodiment............................. 76, 167
emergence of complexity.... 37, 120-22
emergence of consciousness ...37, 120-22, 142
Emergence of Everything, The 120
emotional states...................40, 94-104
emotional survival kit.....................103
emotions 30, 67, 95, 98, 103, 189, 210, 226, 236, 239, 246
   memory and.........................98, 209
Emperor of Scent, The 204
emptiness.............. 43-44, 53, 106, 136
endocrine system......................... 93-95
enlightenment.... 25, 41, 138, 215, 220, 222-23, 231, 233, 237, 240-41, 248
   possibility of................................. 41
enmeshment of body and spirit 20, 45
entrainment 87, 201, 238-39
epiphenomenalism 2
epiphenomenon......5, 123, 129, 171-74, 188, 252, 253
equality of experience, universal .. 159, 181, 248-49, 256
ethics19, 56, 78, 215, 242, 245, 248-49
evolution 4, 12, 14, 15, 16, 17, 25, 30, 35-40, 44-47, 54, 56, 60-62, 66, 70-72, 82, 89-90, 101-03, 107, 109-13, 117-23, 125, 129, 133-34, 137-39, 142, 146, 154-62, 171, 179-82, 184-90, 96-199, 205, 210, 220, 238, 242-48, 254
   bloody in tooth and claw 82, 243
   computer analogy 133
   consciousness, of 17, 184, 210
   creationism and.............. 70, 118-19
   emotions and .... 101 -104
   human mind, recency of............ 158
   model, cube of consciousness ..178-81
   sensorium
      astonishing artifact of...........254
      immense antiquity of............ 161
   theory of......................................117
evolution of consciousness
   elsewhere in the universe 129
evolutionary biology192, 194, 242, 254
   Buddhism and 57

INDEX

excitatory effects 169
existential "qualia"
    angst 40, 246
    despair 246
    isolation 66
existentialism 41
experience 2, 5, 6, 7, 12, 16, 20, 30, 36, 48, 52, 53, 56, 58, 65, 67, 73, 78, 81, 85, 87, 89, 100, 110, 121, 122, 127-29, 136-37, 143, 145-46, 151-56, 159, 161, 169-73, 181, 183, 186, 189, 190, 195, 200, 203, 207, 226, 235-39, 248, 250, 253
    binocular vision, of 151-55
    duality, of 136
    experiencer, bridge between 145, 235, 250
    maps .... *See* maps, experienced
    pleasure and pain, of 125, 249
    time, of 195-97, 201
extrastriate visual cortex 147
extrospection 195
eyespots 68, 72, 119

F

Feeling Buddha, The 216
fetus 47, 179, 180, 187-88
    brainstem and 184-85, 187-88
    consciousness and . 184-85, 187-88
field, electromagnetic .... *See* electromagnetic field
field, visual 6, 88, 126-27, 153, 156, 253
First Sermon of the Buddha, The ... 215
    analysis of 215-30
    text of 231-33
flying buttresses, meninges as 163
FMRI, brain as 5, 87
fossils, transitional 119
Four Noble Truths, the 62
    analysis of 215-30
fractals 173
frequency 203, 205
    brain waves and 97
    neuronal resonators and 3
    sense organs and 124, 159
    sound waves and 174, 205
frontal lobe 16-17, 157
    cortical columns in 157-58, 208
    rational thought and 16-17, 157, 208

Fundamental Neuroscience 161

G

Gaia 118
Gautama 217, 223-27
Gedankenexperiment .... *See* thought experiment
genes 5, 60-61, 109-10, 170, 185-86, 191
genetic
    code 60, 109-10, 170, 184-85, 191
    theory 46-47, 61, 109-10
geometry of three-space 151-52
gestation, consciousness, role during
    .... 184-85, 187-88
gland, brain as 6
Glimpses of Abhidharma 50, 52
God   7, 20, 111, 118, 119, 136, 176, 245, 250
gods 21, 22, 23, 24, 78, 79, 80, 83, 241
Golden Mean, the 7
good and evil 108, 136, 223
grandmother neuron 170
gray matter 144, 145
greed . 25, 107, 108, 112, 221, 224, 226, 230, 231, 245
ground of existence 54, 74, 76, 129, 254
gyrii (gyrus) 145

H

habit energy 60-61
half a soul 120
hallucinatory display. 82, 146, 158, 182
happiness. 35, 40, 41, 58, 62, 106, 107, 210, 220-23, 230, 243, 249, 255
harmonic
    analysis 202
    overtones, missing fundamental
        and 142, 205, 212
    presentation 14, 124, 203, 205
    relationships 3, 7, 95, 97, 124
hate 25, 63, 107, 108, 112, 115, 226, 230
hippocampus 188
Homo aestheticus 198
Homo sapiens 17, 120
homunculus... 7, 14, 16, 84-87, 124-25, 144
hydrodynamics, mindstream and .. 175

INDEX

hypothalamus 188

I

ignorance 25, 26, 35, 39, 50, 54-55, 65-67, 77, 172, 222, 253
illumination 16, 142, 149, 152, 163
illusion ... 2, 7, 12, 17, 31, 34-36, 38, 41, 44, 54, 56, 65 -73, 80, 86, 93, 123, 128-29, 132, 136, 137, 144, 146, 149, 151, 154, 156, 175-76, 193, 194, 196, 200-01, 207, 211, 234, 239, 248, 251, 253
    missing fundamental... 205-06, 212
    painting techniques and 154-55
    time and 200-03
image connectivity 147
IMAX® 73
immediacy 199
impermanence 44, 50, 107, 108
information in DNA 60, 61, 109-10, 112
information processing model 4, 6, 31, 36, 39, 40, 107, 113-14, 121, 125, 134-35, 142, 147-48, 155, 169, 171, 172, 181, 183, 207, 210
inhibitory effects 99, 169
instrument, musical, brain as 6, 25, 40, 205, 210
integration, mind-brain 14, 26, 94, 111
Intelligent Design, Theory of .. 119, 169
intelligent machine, brain as 171
introspection 7, 35, 77, 85, 194, 254
iridescence 68-69
Isipatana 229, 231, 233
isolation, existential 66
I-thou relationship 82
I-thought 36, 111, 136

J

Jesus 56, 111, 240
Jianzhi Sengcan 256
Juhan, Deane 48-49

K

Kali 243, 244
karma 56-57, 59-63, 107, 109
    doctrine of 59
"Keeping Body and Soul Together: A Theory of Neural Resonance" .. 198-99

Kondañña 227, 228, 229, 233

L

laminar organization 146, 208
Lao Tsu 132
lateral geniculate nuclei 155, 156
layers of neurons, six 146, 157
leakage of *Light* 130, 253
LGN 155, 156
liberation 44, 54, 55, 56, 60, 61, 62, 65, 76, 77, 107, 138, 216, 222, 223, 243, 244 and
Light 27, 84, 90, 97, 143, 159
    awareness and 3, 6, 35, 36
    circles of 2, 66
    epiphenomenon, as 5, 128-29
    expansiveness of 143
    fetal development, active presence during, as 37, 184, 186-88
    interpenetration with flesh 163
    leakage of 253
    metaphor, not as 5, 168, 210
    mindstream and 37, 176
    near-death experience of 137-38, 176
    nonduality of 135-38, 142, 168
    patterns of .6, 38, 84, 148, 162, 189, 190, 252
    quantum electromagnetic field and 3, 5, 36, 45, 83, 135, 190
    religion, spirituality and ... 5, 32-33, 45, 129, 132, 135-38, 168, 176
    sculptural medium, as 33
    self, as 36, 70, 71, 81, 186, 211-12, 251
    "spectral color" and 148, 160
    visible . 30, 32, 89, 124, 147, 202-03
light show 4, 15, 126, 164-65, 204
lightbody .. 30, 36-37, 137, 147-49, 151-63, 171-74, 181-84, 187-90, 252
lighting effects 4, 38, 68-71, 87-90, 134, 137, 148, 154, 158, 160-62, 196, 205
Light-pump 181, 184, 188, 195
    activity in fetus 37, 184, 186-88
Light-rail 254
Light-violin .. 6, 31, 89, 174-75, 195-96, 204-05, 210
liking and disliking 27, 40, 210, 253, 255

INDEX

limbic system 35, 47, 48, 181, 182, 185, 188, 203
logic gates............................11, 121, 206
luciferase 160

M

macromolecules 112, 120
magic mirrors...................................128
mahamudra 49, 53, 55
Mahayana...*See* Buddhism, Mahayana
mandala, memory 209
maps..124-29
    auditory 14, 124
    consciousness and ....
    cortical and subcortical ..30, 31, 33, 38, 124, 147, 148
    cybernetic.................... 125, 126, 128
    developmental constraints 125
    experienced, determining whether 155, 160-61
    retinotopic............ 14, 124, 125, 253
    somatosensory 14, 84, 124, 125
    sound frequency..............................
    tonotopic 14, 124
    topographic sensory . 126, 128, 144, 181, 187, 195, 196
    visual ..14, 83, 88, 124, 127-29, 147, 253
Mara 217, 223
marga.................................. 216, 225-34
marquee, theater...................... 2, 9, 38
meditation 5, 41, 43, 49, 50, 53, 62, 97, 167, 215, 217, 229, 230, 234, 235-40, 246-48
    breath 237, 238-40, 246-47
    Buddhist.................66, 215, 238-48
    calm abiding.............................. 247
    counterintuitive aspect of...235-36, 243
    introspection........................ 77, 194
    mindfulness......229, 235, 239, 246, 247
    upon unvarnished reality 41, 243
    yogic, compared........................ 238
medium
    conscious......................................201
    neuronal activity as 4, 5, 30, 33, 36, 71, 148, 169, 171, 172, 175, 182, 187, 188, 201, 252
    passive, nerve impulses as... 169-75

propagation of *Light,* for ... 4, 5, 30, 33, 36, 71, 148, 169, 171, 172, 175, 182, 187, 188, 201, 252
    sculptural, *Light* as 33
    transparent liquid 163-64, 188, 252
melanin.. 69
memes, cultural DNA, as 61
memory .. 12, 16, 36, 71, 72, 83, 84, 95, 96, 98-100, 103, 128, 137, 155, 157, 158, 170, 182, 195, 197, 201, 203, 207-11, 241, 244
    mandala 209
    palaces of...................................208
    personal, in microgenesis 182
    sense 203, 209
Mendel, Gregor 61
meninges 144, 163, 165, 188
    flying butresses, as.....................163
    role in sensorium 163
    shock absorber, as.............. 163, 165
mental space......................................154
metaphysical leap............................ 188
microgenesis...35, 36, 53, 123, 180-84, 192, 193, 195, 197, 199, 254
microgeny
    model, cube of consciousness... 180
microtubules 190-91
Middle Way, the 228, 231, 242
mind
    analytical, calming 237
    appropriate use of 246
    awakened..61
    Buddhist view of... 56, 157, 199, 207
    evolution, recency of 158
    insubstantiality of 53, 158
    perceptual field of (internal) 157-58
    presentation, haziness of 158
    sense organ, as 157-58
mind only 53, 56, 57, 58
mind-body
    connection ... 72, 2, 81, 94, 110, 113-14, 128-29, 132, 144, 174-76, 190, 196-97, 254
    identity of, apparent 84, 87, 176
    interpenetration 163
    nonduality and 132, 135-37
mindfulness..... 223, 228-31, 234, 235, 236, 238, 239, 241, 246, 247, 254
mindscape 158
missing fundamental 142, 205, 212

modern synthesis, the 117
moksha 81
molecules of emotion 94, 94-104, 210, 249
Molecules of Emotion, The 94
molecules, self-replicating 109, 112, 129
moods 98-104
    evolution of 101-04
    memory and 98, 210
    survival mechanism, as 103
Morowitz, Herald J. 120
multiplex, cerebral, the 126
music and the brain.... *See* instrument, musical, brain as
musical intervals 31, 173
musical mathematics 202
musical pitch .4, 124, 148, 174, 202-03
myelin 144
mystics 5, 8, 17, 21, 111, 135-36, 235

N

natural selection .... *See* evolution
nature vs. nurture 191
near-death experiences .... 137-38 , 176
neo-Cartesian nondualism. 14, 15, 135, 149
neo-Cartesian theater 15, 149
neo-Darwinism 117, 119
neocortex 35, 47, 48, 53, 157, 162
nerve fibers 144
nerve impulses, passivity of 4, 148, 171
neural resonance 4, 25, 40, 89, 101, 198, 205, 253
neural resonance chamber 9, 31, 41, 101, 173, 195, 196, 197, 202, 207, 210, 253
neurobiology 1, 2, 4, 93, 121
neurobioluminescence 5, 12, 29, 37, 56, 71, 75, 87, 89, 90, 123, 132, 138, 148, 155, 157, 160, 167, 171, 174, 242, 244, 248, 250, 252
neuroendocrinology 93
neuroimmunology 93
neuromodulators 99, 101
neuronal
    continuum 124
    oscillations, attention and 239
    oscillators. 3, 4, 13, 31, 88, 173, 195, 202, 210

proximity 38, 124, 125, 147, 148
resonance .100, 197, 206-07, 210-11
    sensation, memory, analytical thought and 207-10
resonators 31, 89, 173, 195, 197, 202, 206
neurons
    binary on/off switches, as7, 13, 206
    circle of, fundamental unit, as to, 3, 4, 148, 206
    pacemaker 38, 48
    passivity of 4, 148, 169, 172, 175, 195
neuropeptides .46, 94-97, 99-102, 122, 165, 170, 172, 185, 191, 196, 249, 253
neurotransmitters 11, 94, 99, 100, 135, 165
nirodha 216, 225-33
nirvana 54, 64, 107, 136, 219, 230
nodes and antinodes 174
nonaction 136, 235
nonbeing 136
nonconsciousness ... 39, 71, 113-14, 121, 122, 123, 130, 155, 169, 172, 201, 253
    brain, physical, and 122, 172, 197
    neurons and 148-49, 170-71, 253
    presensorium and ... 40, 71, 122, 197
nonduality ... 132, 135-37, 140, 143, 149
noninterference 171
nonlinear systems 190
no-self 45-46, 50, 66
not two 132, 136
now, the 48, 201, 211, 246
now-realization 49

O

observer 36, 69, 155, 184
occipital lobe 14, 15, 16, 17, 86, 124, 166
ocelli 68
ocular dominance columns 147, 151-52
ontogenesis 47, 53, 181, 186
ontogeny 46, 179
    model, cube of consciousness ... 179
ontology 120, 123, 193
orientation columns 147, 162
Origin of Species, The. 44, 61, 119, 226
    Buddhism and 44

INDEX

orthodox view, Buddhist 217, 223, 227
oscillations, time and............... 200-05
oscillators ...... *See* neuronal oscillators

P

painting techniques 154-55
pairings, neuronal presentation structures and qualia perceived 160-61
palaces of memory 208, 241
Pali... 215, 228
paradox, time as.............................. 201
paramitas 248
paranoia 53
parasitism.. 73
parents............... 102, 122, 179, 191, 199
parietal lobe 15-17, 48, 84, 166
parinirvana 219
passivity of nerve impulses 4, 148, 169, 172, 175, 195
Path, the .. 44, 55, 60, 65, 76, 219, 227, 229, 230, 232, 234, 235, 254
paticca-samuppada 217, 221, 222
Patthana 199
peacock............. 67, 68, 69, 70, 72, 160
perception 81, 96, 100, 101, 118, 126, 137, 142, 143, 147, 148, 157, 162, 181, 183, 184, 192, 205, 207, 221, 226, 236, 246, 247, 248
Persian carpet 7, 144, 145, 146, 159, 162, 252
personality, persistence of.............. 197
perspective, geometrical................. 154
Pert, Candace 94, 95
pessimism, Buddhism and 106
philosophy, Buddhist..... 49, 57, 60, 61, 66, 107, 216, 234
photogenesis 36, 37
photons... 3, 6, 15, 29, 32, 34, 126, 134, 160, 253
phylogenesis.. 46, 47, 53, 125, 179, 181, 186, 203
phylogeny... 46
  model, cube of consciousness ... 179
physics............ 33, 43, 45, 98, 135, 250
  role in East-West Dialogue 43
  theoretical 29, 43, 240
pia mater 12, 144, 145, 146, 157, 163, 164, 252
  role in sensorium 163

thinness of 163
transparency of 163, 252
pixellation.... 13, 32, 134, 145, 157, 158, 252
pixels. 6, 9, 12, 13, 32, 68, 70, 146, 155, 160, 163
plane of focus 152, 153, 154, 155
planes of consciousness .. 52, 180, 182, 186, 194
planetarium, personal.................... 166
pneumatic tubes, cerebral blood vessels as 164
poem 145
point-by-point correspondence..... 124
postulate, metaphysical 29, 32, 40
practice, spiritual 41, 49, 50, 55, 62, 66, 72, 76, 111, 112, 143, 193, 216, 220, 223, 225-27, 229-30, 235-40, 242-43, 246-48
preobject................................... 48, 182
prebiotic era 112
preconscious animals... 16, 39, 65, 123, 128-29
predator-prey relationships 66, 67, 70, 72, 102, 103, 111, 112, 113-14, 122, 129, 204, 226
predicament of the soul 25, 45
preferred embodiment........... 138, 160
prefigurement, arguments against. 39, 123
presentation 9, 12, 13, 15, 17, 30-34, 48-49, 71-73, 81-90, 122, 134, 137, 142-49, 154-62, 166, 168, 170, 172, 189, 196, 211, 226, 249, 252, 253
  five senses, robustness of........... 158
  mind, of 157
  mind's, haziness of..................... 158
  neurons.... 9, 13, 29, 31, 89, 97, 146, 157, 160, 170, 252
  orientation columns and............ 163
  plane of 4, 161
  qualia, of 145-46, 159, 160, 203
  screens 31, 33, 34, 38, 147, 149, 156, 253
  verisimilitude of 14, 38, 73, 89, 128-29, 134, 137, 161, 183
proprioception 84, 145, 183
proto-awareness........................ 29, 37
proto-consciousness 30
proto-mood 103

265

INDEX

proto-self 181
Psyche
   beauty and 24
   bewilderment of 73, 81, 128, 175
   contrasted with Buddhist view ... 50
   Cupid
      at mercy of ... 80, 81, 93, 101, 110
      identification with 83, 84
      marriage to (carnal body) 21, 73, 80, 81, 103-04, 199
   curiosity of 23, 26, 79
   emotions and 96, 101, 110-11
   hell realms, in 77, 79-82, 111, 251
   journey of awakening of 77, 251
   naïveté of 196
   psychein (to breathe), from 24
   quandary of 25
   reality of experience of 81-83
   sensorium consciousness, as personification of 24
   sentimentality regarding 50, 81
   soul, as personification of 24
   universal incarnation of 53, 83, 254
   virtual 196
   wraparound world of 82, 84
Psyche's palace ... 15, 17, 22, 23, 25, 26, 80, 101, 110
psychology, Buddhist 52, 230

Q

qualia .. 145-46, 153, 155, 159, 160, 161, 166, 203
quale .... *See* qualia
Quan Yin 230, 243, 244, 245
Quantum Brain, The 190
quantum field
   consciousness and . 5, 14, 15, 19, 29, 30, 32, 33, 40, 119, 128, 205
quantum mechanics 39, 43
quantum paradox 43

R

Ramana Maharshi 111, 136
Rational Biology 117
ratios
   aesthetics and ... 3, 4, 7, 31, 173, 175, 202, 206, 211
   neural resonators and ... 3, 4, 31, 84
reactivity 112, 196, 220, 223, 226, 229, 230, 234, 236

reality 5, 6, 8, 11, 14, 18, 33, 40, 41, 49, 51, 52, 56, 57, 65, 66, 70, 80, 101, 104, 107, 120, 135, 136, 137, 138, 142, 146, 149, 151, 157, 161, 176, 193, 194, 201, 236, 242, 243, 248
Realmeditazion 41, 243
reductionism 1, 71, 118
reincarnation 49, 59, 60, 187, 221
relationship of body and soul .. 4, 6, 26, 198
religion and science .... *See* science and religion
REM sleep 97, 180
replication of DNA 110, 112
res cogitans and *res extensa* 8, 135
resonance ...... *See* neural and neuronal resonance
   patterns 84, 96, 97, 98, 173, 174, 189, 190, 210, 238
   time and 200-07
   universal phenomenon, as 198
resonators
   chaotic 98, 172, 173, 174, 177
   oscillating neuronal 3, 211
Response to Afflictions, the 218
retina 14, 30, 124, 125, 126, 156, 170
retinotopic map *See* maps, retinotopic
retrograde consciousness 86
reverse orientation .. 15-16, 18, 86, 254
rhythms 3, 7, 142, 148, 198, 205
Right View, Thought, Speech, etc. 228
rings of neurons 3, 4, 31, 88, 211
RNA 112

S

sadayatana 246
samadhi . 223, 225, 228, 229, 231, 237
samatha 148
samsara 54, 65, 107-08, 136, 226, 230
samudaya 216, 218-34
sangha 76, 242
Satinover, Jeffrey 190
Schrödinger's Cat 45
science and religion ... 8, 19-20, 29, 43, 44, 49, 57, 65-66, 71, 129, 135, 237, 242, 243, 244
screens, presentation .. 6, 13, 14, 15, 30, 31, 32, 33, 34, 38, 69, 73, 74, 81, 90, 100, 126, 128-29, 134, 147, 149, 152, 155, 156, 160, 253

INDEX

self
- animals and (selfhood in animals) 39, 40, 65, 248-49
- center of presentation. 81, 132, 156, 246, 248-49, 253
- concept 41, 184, 239, 246, 248
- elucidation of, Buddha's 44
- emergence of 53-55, 121-22, 195
- fabrication, as 45
- illusory nature of 41, 128-29, 195, 241, 255, 256
- inference, as 12, 83
- LGN, projected in 156
- microgenesis and 36, 41, 48, 49, 182, 196
- nature, emptiness of 44, 106
- neurobioluminescence and 71
- other, and 54, 82, 111, 136, 248
- persistence of 199
- personality and 197
- physicality of 84
- refuge, as 44, 54
- sensorium and 11, 35, 71, 82, 156
- sentient beings, all, and 44, 248-49
- separately existing 50, 71, 84, 93, 226, 239, 248
- solipsism and 66, 128-29
- spiritual inflation and 248
- spiritual practice and 111, 239
- welcoming 41, 176, 197, 200
- world, and .. 34, 36, 37, 49, 86, 136, 137, 141, 168, 182, 184, 246

Self, the 136, 193

Self-Embodying Mind, The 35, 47, 48, 141, 181, 193

self-inquiry 127-28

Selfish Gene, The 110

sensation, consciousness as pure ... 162

sense memories 203, 209

sense organs ... 4, 25, 31, 44, 45, 65, 66, 67, 71, 113-14, 122, 125, 157, 159, 160, 161, 203, 207, 226, 229, 253

senses, mind as one of six ... 54, 55, 83, 110, 128, 137, 202, 209, 226, 241, 246

sensorium
- animal brains and 161, 210, 248-49
- apparent reality of 83
- artifact of evolution, most astonishing, as 254
- blood brain barrier, role in 164
- brain structures and .. 4, 83, 87, 144
- consciousness 18, 37, 39, 40, 56, 107, 120-23, 149, 174, 178-81, 215, 248, 252, 254
- dark half of 18
- display, generating the 13, 30, 97
- emotions and 95-104
- evolution of. 40, 66, 90, 113-14, 119, 122-29, 161, 169, 196, 197, 215, 248, 252
- evolutionary antiquity of ... 123, 161
- extravaganza 4, 9, 134, 252
- forward half of 15, 17
- fully elaborated 34
- illusory nature of 65, 72, 82, 93
- information, distinct from ... 169-70
- introjection of *self* into 31
- *Light*-violin, as ...... *See Light*-violin
- lightbody and 36, 173
- meditation and 239
- meninges as cytoarchitectural feature of 163
- miraculous simultaneity, as 30
- neurons' firings and 36, 38, 134, 147-48
- neuropeptides and 95, 100
- objective presentation, as 81
- presentation 37, 39, 89, 137, 146, 207, 211
- quantum electromagnetic field and 83, 170-71
- refinement of 46, 73, 161
- refuge as 76
- reversed orientation of 15-16, 18, 86, 254
- spherical form of the presentation 12, 93, 211-12, 246
- subjective experience of 96, 110, 152
- subjectively centered 12, 16, 31, 246
- theater of the mind 25
- thoughts, presentation of 157-59
- transition to 119-22
- *self* and world surround and 35, 36, 84, 211-12, 246
- special effects and 30, 87
- three dimensional space in .. 152-56
- verisimilitude of 14, 38, 73, 89, 128-29, 134, 137, 161, 183
- vision's primacy in 32, 151-56

sentient beings 7, 40, 45, 46, 61, 72,

267

sentient beings (contd.) ..74, 100, 104, 107, 118, 128-29, 159, 181, 198, 248, 253
    definition of 65-66, 71
    saving all 64, 109, 136, 220, 243, 245
sentimentality......... 41, 50, 81, 110, 111
sfumato 154
Shakti and Shiva 254
six sense gates, the 246
skandhas 55, 241
sleep 88, 97, 180, 182, 186
    consciousness during 186
        maintenance tasks and 186
    fetal development and 186
somatosensory cortex 84-87
soul... 1, 4, 5, 6, 7, 8, 21, 24, 25, 26, 38, 45-46, 50, 70, 73, 77, 104, 119, 120, 146, 149, 155, 179, 187, 188, 189, 198, 199, 244, 245, 254
special effects 68, 87, 137, 154, 158, 161
special status, humans' unwarranted 248-49
spectrum
    electromagnetic 6, 32, 71, 89, 145
    visible light, of 6, 32-33
Sphinx, Riddle of the........................ 20
spontaneous biophoton emissions 160
spontaneous generation 121
Squire, Larry R. 161
stage illusions 137
stereopsis 124
straw man arguments 119
striate cortex .... *See* visual cortex
String and Superstring Theories 43, 204
structural optics 69, 70-71
structured electromagnetism 207
stuff of consciousness 171, 183
subject/object, nonduality of ... 36, 111, 135-37, 169, 252
suffering *(dukkha)* 217-33
Sufism 5, 135-36, 235
sulci (sulcus) 16, 145
Sutra of the Turning of the Wheel of the Dharma, The .... *See* Turning of the Wheel of the Dharma Sutra, The
Sutra on the Full Awareness of Breathing, The 237

sutras 55, 56, 111, 215-30, 33-35
symbiosis 73
sympathetic vibration 174, 201, 203, 211
synesthesia 146

T

T'ai Chi Ch'uan 143-44
taking refuge 76
Tantra 55, 132
Tao Te Ching 132
Taoism 43, 132, 135, 136, 137
Technicolor display 30, 128
television screen 6, 32, 69, 127, 149, 155
temporal lobe 14, 16, 85, 166, 182
terminology (word sense disambigation) 14, 19, 24, 32, 33, 50, 56, 57-58, 65-66, 71, 83, 120, 121, 126, 142-43, 220, 222, 223, 227, 228, 230, 241, 256
thalamus 85, 155, 188, 238, 239
thangka paintings 108, 209
theater of the mind ... 11, 15, 25, 73, 74, 90, 134, 149, 154, 156
theta brain waves 97
thin-film interference 69
thought experiment 7, 45, 95, 122, 127-28
thought-moment 37, 123, 193, 197, 200
thread
    Persian carpet 13, 146, 165
    sutra, etymology of..... 131, 215, 243
Three Marks of Existence, the .. 44, 106
Three Poisons, the 25, 106-09, 112
Three Treasures, the 76
three-dimensional space .. 5, 6, 30, 83, 144, 149, 151-56, 162, 165-66, 168, 179, 183, 207, 209, 253
three-space, experiencing a two-dimensional image in 153-54
Tibetan Book of the Dead, The 138
time .. 3, 5, 6, 33, 53, 57, 59, 72, 77, 87, 90, 99-100, 102, 108, 134, 136, 138, 169, 180, 186, 187, 188, 195-205, 211, 215, 237, 244, 253
    counting and 202-03
    invention of 200
    perception and 195-207

INDEX

perception of 200-03
resonance and 200-06
topographic sensory maps... *See* maps, topographic sensory
trabeculae..163
transition to sensorium consciousness 40, 107, 113-14, 119-20, 122-24, 197
transparency of brain fluids 163-65, 188, 252
transparency of *pia mater*. 12, 144-46, 157, 163-64, 252
Trungpa, Chögyam .. 50, 52, 53, 54, 55
tuning forks, sense organs as 203
Turin, Luca..................................... 204
Turning of the Wheel of the Dharma Sutra, The .. 215, 216, 219, 222, 223, 234, 242
    analysis of............................. 214-30
    text of..................................... 231-33
Tuszynski microtubule module 190-91
Twelvefold Chain of Causation, The 57, 108, 221, 222
two-dimensional surface
    cerebral cortex as .. 14, 33, 124, 125, 153, 252, 253
    access by consciousness, for.162
    surface area maximized 144, 254
    retina as.....14, 30, 124-26, 156, 253

U

unconsciousness 21, 54, 74, 79, 101, 158, 171, 186, 189, 238
unhappiness......................................107
unorthodox Buddhist views 56-62, 217-33
unsatisfactoriness 40, 44, 106, 226, 230
Ussher, James 118

V

Vajrayana ... *See* Buddhism, Vajrayana
Veblen, Thorsten...............................67
ventricles...188
verisimilitude of sensory presentation 14, 38, 73, 89, 128-29, 134, 137, 161, 183
viability, consciousness and 187-88
vibrating universe 198, 204
vibration 6, 89, 147, 173, 175, 199

sense organs and .. 89, 172-73, 200-06
time and..............................200-06
violin, *Light*-............... *See Light*-violin
virtual Psyches 196
visual cortex ...6, 14, 15, 16, 30, 32, 87-88, 124-27, 147, 149, 151-56, 159-63, 170, 189, 203
    auxiliary visual cortex and... 14, 147
    binocular vision and 124, 151-55
    evolutionary constraints upon 14-16
    incurvate structure of. 124, 130, 156
    posterior location of.... 14-15, 17, 30
    retinotopic map, as 14, 124, 125, 253
    transitional structure, as 40, 123-24
voltage, extravagant ...3, 32-33, 37, 83, 252

W

wastefulness of calories 134
watchmaker and the watch, the...... 119
watts, power output of brain 13, 32
wavelength color 89, 146, 154, 160
Weltanschauung 2
wetware 121
Wheel of Becoming, the .. 108, 112, 123
wheel of cyclical existence 25, 108, 226
will
    breath meditation and the 235, 237-38
    chaos and the 211
    conscious, the 72, 188-91
    egoic, release of the 235
    existence of the, proof of the 188-91
    obstacle as 235
wiring, axonal-dendritic 3, 5-6, 36-38, 99, 146-48, 169-70, 185
wisdom ... 24, 40, 54, 61, 118, 136, 216, 222, 232, 237, 242, 248
word sense disambiguation ....*See* terminology
wu wei 136
Wumen Huikai 247

X Y Z

Xin Xin Ming.............................255-56
Yama 108
Zen .... *See* Buddhism, Zen